Statistics

a dictionary of terms and ideas

GW00497096

Arrow Reference Series

General Editor: Chris Cook

Other books in the series:

Astronomy
A dictionary of space and the universe
Iain Nicolson

Basic Economics
A dictionary of terms, concepts and ideas
Tim Congdon and Douglas McWilliams

Computing
A dictionary of terms, concepts and ideas
Anthony Hyman

Earth Resources
A dictionary of terms and concepts
*David Dineley, Donald Hawkes, Paul Hancock
and Brian Williams*

The Environment
A dictionary of the world about us
Geoffrey Holister and Andrew Porteous

Photography
A dictionary of photographers, terms and techniques
Jorge Lewinski

Statistics

a dictionary of terms and ideas

Colm O'Muircheartaigh
and
David Pitt Francis

Arrow Books

Arrow Books Limited
3 Fitzroy Square, London W1P 6JD

An imprint of the Hutchinson Publishing Group

London Melbourne Sydney Auckland
Wellington Johannesburg and agencies
throughout the world

First published 1981

© Colm O'Muircheartaigh and David Pitt Francis 1981
Illustrations © Arrow Books Limited 1981

Set in VIP Times by BSC Typesetting Ltd

Made and printed in Great Britain
by The Anchor Press Ltd,
Tiptree, Essex

ISBN 0 09 920330 8

Contents

Using this dictionary

Not all the terms defined in this dictionary are cross-referenced to one another each time they occur. Only when the understanding of a term used in an entry adds to the reader's comprehension of the particular area under discussion will the term be marked in small capitals, thus:

Absolute scale.   A scale of MEASUREMENT where the scale scores may not be modified in any way without destroying the scale.

A single arrow has been used for 'See'; double arrows for 'See also'.

A guide to further reading may be found at the back of the book.

Introduction

Statistics is one of the most fundamental of all the sciences. No human decision can be made without involving statistical reasoning at some level. Statistics deals with uncertainty, with chances, with probabilities. The words 'likely' (and 'unlikely'), 'expected', 'probable', 'favourite' – in horse-racing and other sports, 'upset', 'surprise' all denote uncertainty and permeate our entire existence.

As a science, statistics occupies an unusual position in two respects. First, although it deals with empirical data, there is no field in which data are exclusively 'statistical'; rather, the statistician works with data from other subject-fields. Second, all of statistics is based on chance events, and the science of statistics is concerned with exploiting the regularities which arise from the operation of chance on a large number of elements or individuals simultaneously, or from the repeated occurrence of chance events.

The definition of statistics has occupied the attention of statisticians for many years. By 1935 over 100 definitions had already been formulated. Kish (1978) suggested a broad outline of the subject – 'Statisticians and statistics deal with the effects of chance events on the treatment of empirical data' – which, though controversial, seems to us to embody the basic attitude of mind that lies behind all statistical operations. There are thus three basic components of the subject: (1) chance, or probability, which encompasses the fluctuations or variability occurring in nature; (2) empirical data – the statistician's raw material – which can be taken from any or all fields of human activity; (3) the way in which the chance element is taken into account in assessing the data.

A number of examples here may help to illustrate the breadth and the cohesiveness of statistics. Censuses, particularly censuses of population and economic activity, have long been an important component of the subject. Their statistical importance, however, is in the fact that the way in which the data are presented and analysed recognizes the influence of variability (or its counterpart, regularity). It is this acceptance of the role of chance which distinguishes the statistician's work from that of accountants, book-keepers and the takers of inventories. Indeed, the traditional approach to accounting – concerned as it was with stewardship and completeness of coverage – has in recent years, developed and come to rely more and more on statistical techniques, for example in sampling for auditing and production control.

Another basic strand in the development of statistics has been gambling. In the seventeenth century, some of the prominent gamblers of the day approached the mathematicians with sets of problems on the calculation of the odds or chances of particular combinations of numbers in dice games. The questions themselves suggest enormous experience of dice throwing since the differences in probabilities between the events concerned were sometimes lower than one in 100, which would require as many as 10 000 throws to establish empirically.

An interesting development in gambling is illustrated by the Premium Bonds scheme where the gambler's initial investment (the purchase price of the bond) is at all times protected and available for withdrawal. Here it is the interest on the investment which is the gambling stake so that the system provides a nice balance between thrift and adventure.

One of the earliest and consistently most successful applications of statistics (or area of statistics) is the field of life assurance. Here the individual is in effect offered the choice – in financial terms – between his own individual future and some average or typical future. The same applies in health insurance (i.e. against sickness) and fire insurance, for example. Thus the individual chooses to pay some amount of money on the condition that if some event occurs to him he will be compensated, whereas if the event does not occur to him he receives no return on his investment. The process is therefore, for the individual, simply substituting small frequent expenditures for the possibility of either no expenditure or a very large expenditure. It is a transfer of risk from the individual to the aggregate of individuals.

One other area which warrants inclusion here is sampling, where the chance element is introduced into the system by the statistician. Thus, rather than measure some characteristic of the whole population, the statistician decides to restrict himself to a small part of the population and to use the principles of inference to make statements about the whole based on the observation of the part. All human experience is based on reaching conclusions based on a sample of some sort. For example, whether one likes or dislikes a new acquaintance may depend on behaviour at first meeting, although such behaviour is only a tiny sample of the individual's total behaviour over a lifetime. Similarly whether we decide a book is worth reading may depend on our reaction to the first few pages; we may choose a magazine for a train journey on the basis of a quick flip through the pages.

Scientific sampling – defined in the text under 'probability sampling' – is simply a set of systematic and rigorous procedures whereby we can improve our decisions by removing certain biases from our choice of sample.

Another major field, related to that above, is that of prediction. Here the unknown or unobserved component is the future. The observations from the past and present, together with postulated or observed regularities, are used to predict future occurrences.

In this book we have concentrated our attention on the general principles and ideas of statistics and illustrated these ideas in the context of particular applications. Many of the examples are related to business and accountancy, others to the area of survey research and opinion polling, and others to econometrics. There are areas of application on which we have touched only lightly – population studies, or demography, is one such area. The emphasis is on the basic ideas and on terms frequently found in many diverse fields of interest. We do not discuss in detail the myriad statistical techniques available, nor do we try to provide a manual on statistical manipulation. Our aim is to give an overall view of the more important ideas and an explanation in broadly non-technical language of the content of the subject and to illustrate these by providing simple examples of the kind of application for which they are useful.

A

a (as regression constant, and as coefficient). In some texts where statistics is taught in association with business studies, economics, management accounting and econometrics the letter *a* is sometimes used to indicate the REGRESSION CONSTANT, that is, the non-varying term in an equation expressing the relationship between a dependent variable and one or more independent variates.

For example, in accounting the relationship between output (or quantity produced) and cost is often expressed as a simple LINEAR REGRESSION equation $Y = a + bX$, where Y represents total cost, X represents output, b represents cost per unit and a indicates overhead, or fixed cost of production, the aggregate of those items such as rent and administrative salaries which, between well-defined limits, do not vary with production. In basic economics a similar bivariate equation is also used to define the relationship between national income and production. Econometrics uses the term in multivariate models and in multiple regression models such as

$$Y = a + b_1x_1 + b_2x_2 + b_3x_3 + \ldots + b_nx_n + U$$

In mathematical statistics, *a* is frequently replaced by β_0, b_n by β_n, and U by ε, so that the above multivariate equation would be written

$$Y = \beta_0 + \beta_1x_1 + \beta_2x_2 + \beta_3x_3 + \ldots + \beta_nx_n + \varepsilon$$

The relationship between two variates in a bivariate equation can be expressed diagrammatically (using the cost/output illustration given earlier) in the following way:

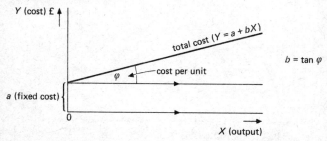

Another, and unrelated, use of *a* is as the indicator of coefficients in LINEAR PROGRAMMING. For example, Maximise $a_{11}x_1 + a_{12}x_2 \ldots$ subject to a number of constraints. ($\blacklozenge$ LINEAR PROGRAMMING.) These two uses should not be confused.

Abnormal curve, abnormality.   The term abnormality should not be confused with SKEWNESS, the latter term meaning positive or negative asymmetry in a distribution curve. Abnormality is more extensive. Any frequency curve may be described as abnormal if it differs from a NORMAL CURVE which can be drawn, having the same median and dispersion as itself.

Abnormal curves may be simple, complex, bilateral, unilateral, uniform and disform. A simple abnormal curve is one in which each of the two branches would cross the corresponding normal curve only once: whereas a complex abnormal curve is such that one or both of its branches cross the branches of a corresponding normal curve more than once. When curves exhibit abnormalities on only one side, or on both sides, they are said to be unilateral, or bilateral, respectively. Uniform abnormality is the occurrence of the same abnormality on both sides (or branches) of the curve, one side mirroring the other side, whereas disform abnormality is the occurrence of different types of abnormality in each branch of the curve.

The following diagrams of abnormal curves are:

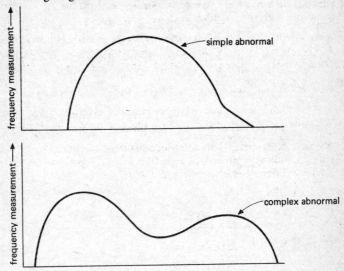

These can be compared with the NORMAL CURVE.

Abscissa. The base line, or horizontal axis, of a graph or polygon. In the case of the FREQUENCY DISTRIBUTION below, for example, the vertical side (or axis) of the graph is used to measure frequency of occurrence of each score (or quantity of a variate) and the horizontal side (or axis) depicts the measurements of quantity, score or size of the variate whose frequency is being illustrated.

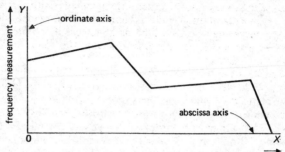

The vertical (frequency) axis in this case is the ORDINATE, and the horizontal (score) axis is the abscissa.

Absolute deviation. The difference between any given value of a variate and a fixed value or constant, irrespective of sign. For example, the MEAN DEVIATION is the sum of absolute deviations from the mean (or median) divided by the number of values.

Example The mean of the series 1, 5, 9, 12, 18, is 9. Actual deviations are -8, -4, 0, $+3$, $+9$. Absolute deviations are simply 8, 4, 0, 3, 9. The symbol $|\ \ |$ is frequently used in formulae where absolute deviations are required. For example, the formula for mean deviation

$$\frac{\Sigma\ |x_i - \bar{x}|}{n}$$

means the sum Σ of each absolute deviation from the mean $|x_i - \bar{x}|$ divided by the number of values (or observations).

Absolute difference. The difference between two quantities or values x and y, taken regardless of sign (i.e. always a positive value). The absolute difference between x and y is written as:

$$|x - y|$$

   (i) If $x = 5, y = 2$ then $|5 - 2| = |3| = 3$
  (ii) If $x = 2, y = 4$ then $|2 - 4| = |-2| = 2$

The actual difference in each case would be:

   (i) $x - y = 5 - 2 = 3$
  (ii) $x - y = 2 - 4 = -2$

Absolute error.   The difference taken as positive between an observation and its true value, hence the extent of inaccuracy between an observation, as expressed in quantifiable terms, and its 'correct' value. This error, or deviation, between an observed value and its true value is also absolute in the sense of not having a sign, i.e. both positive and negative measures being expressed as though they were positive.

Absolute frequency.   The actual number of times a particular measure of a variate or a particular attribute occurs, as distinct from RELATIVE FREQUENCY, its occurrence as a proportion of total frequency. For example, if a school contains 1000 students of whom 200 are 'sixth formers', the figure 200 is an absolute frequency as distinct from the relative frequency of 0.2.

Absolute scale.   A scale of MEASUREMENT where the scale scores may not be modified in any way without destroying the scale. Counting, viewed as measurement, is an example. This is the highest level of measurement but is not particularly useful nor is it commonly found in practical work.

Absolute value.   The value of a quantity taken regardless of sign (i.e. assuming a positive sign). The absolute value is indicated by enclosing the quantity (or deviation or variable) between two vertical lines. Thus the absolute value of x is written as $|x|$.

The absolute value of -3 is $|-3| = 3$
The absolute value of 3 is $|3| = 3$
The absolute value of $2-7$ is $|2-7| = |-5| = 5$

'Absolute' is occasionally used more loosely in contrast to 'relative', e.g. absolute frequency v. relative frequency.

Acceptable Quality Level (AQL).   A simple form of quality control measure. The consumer determines the 'required' proportion of good units (P), and this proportion is termed the acceptable quality level of the batch. The complement $(1 - P)$ represents the proportion of units of defective type which the consumer is able (or prepared) to tolerate.

Acceptance error.   An error which results from accepting a hypothesis, when *the* alternative or *an* alternative is true.

Acceptance region.   A bounded section or region of the EVENT SPACE or SAMPLE SPACE, such that, if the sample point falls within the region, the hypothesis being tested is accepted.

Acceptance sampling.   This term denotes a type of sampling used both in production control and in the control of the accuracy of clerical work. The basic problems are those of deciding the level of confidence at which the audit or inspection department will operate and the maximum possible number of errors (or rate of errors) at that level of confidence.

Using a particular sample size, previous experience may show that more than x per cent of defectives is particularly bad, and tables, using calculations from the Poisson distribution, have been constructed to give for every lot size and ALLOWABLE DEFECTS two factors, n the sample size and c the acceptance number.

Let us assume that for a stated lot size and required maximum percentage of defectives (x per cent) the sample size (n) is 200 and the acceptance number (c) is 2. Application of acceptance sampling would involve rejecting a batch if it contained more than two defective items (e.g. products or records) but accepting it if it contained 2 or less. ◆◆ ALLOWABLE DEFECTS, QUALITY CONTROL.

Accuracy and approximation in computation.   The terms accuracy and approximation are complements or converses of each other. Accuracy usually means the nearness of computations, estimates or published figures to true values. Where figures are in discrete units they may be stated accurately as an exact number of units, 54 men, 226 households, 21 941 library books, etc. But if physical data are used and quantities are measured such measurements may have to be stated to the nearest centimetre, tonne, kilogram or litre. Most physical measurements of size, weight, quantity, etc. are not entirely exact but rounded to the nearest unit of measurement, so as to comply with a legal, recommended, colloquial or contextual standard.

The degree of accuracy (or approximation) can be expressed in the following ways:

1. simply state the number to the nearest thousand, hundred, whole number or decimal place; thus 3752.38 is written as 4000 to the nearest thousand, but as 3752.4 to the nearest decimal place;

2. use $\pm$ to show the degree of approximation or accuracy, so that the above would be expressed as 4000 ± 500;

3. use a percentage to indicate relative approximation or error, and express the above example as 4000 ± 12.5 per cent; or

4. express the degree of accuracy or approximation to significant figures. In this case, 3752.38 would be expressed as 4000 to one significant figure, 3752 to four significant figures and 3752.38 to six significant figures.

Cases 2 and 3 are two methods of expressing the degree of error (or approximation). Case 2 tells us that the maximum absolute error in either direction is 500 while case 3 shows the maximum relative error to be 12.5 per cent.

Addition principle (in probability).   The 'addition principle' forms one of the bases of elementary probability theory. The rule states that, if a set of events are mutually exclusive, the probability of one or other of them happening is the sum of their individual probabilities of occurrence.

The most frequently used illustrative example concerns the casting of a six-sided die. Only one side can fall uppermost at a given throw, so that the probability of a 1 being cast (written $P(1)$), and that of a 2 being cast (written $P(2)$) are mutually exclusive. If the die is unbiased the probabilities of each of its six sides falling uppermost is 1/6, for there is a certainty (probability = 1) that one of its six sides will fall uppermost and, as the die has six equi-probable sides, each side has a 1/6 chance.

The probability that either a 1 or a 2 will occur (as distinct from that of either 3, 4, 5 or 6 occurring) can be obtained by adding together $P(1)$ (i.e. 1/6) and $P(2)$ (also 1/6). Thus the probability of *either* 1 *or* 2 is $P(1) + P(2)$, in this example 1/3. Similarly the probability that 1 or 2 will not occur – that is, that 3, 4, 5 or 6 will occur if the die is cast, may be derived either by adding together the equal probabilities of 3, 4, 5 and 6, $P(3) + P(4) + P(5) + P(6) = 1/6 + 1/6 + 1/6 + 1/6 = 2/3$, or by deducting the probability of either 1 or 2 being cast (1/3) from certainty ($P = 1$), i.e. $1 - 1/3 = 2/3$.

The rule can be applied to non-equal probabilities. For example, a survey shows that of 600 firms, 60 increase market share, 480 maintain market share, 50 reduce their market share and 10 go bankrupt. The probabilities of a typical firm undergoing these events in a specified time are respectively $P(A) = 10$ per cent, $P(B) = 80$ per cent, $P(C) = 8.3$ per cent and $P(D) = 1.7$ per cent. Thus the probability of the typical firm not going bankrupt would be $P(A) + P(B) + P(C) = 98.3$ per cent, or conversely $1 - P(D) = 98.3$ per cent. The probability of its not simply maintaining market share would be $P(A) + P(C) + P(D) = 20$ per cent.

If the events A and B are not mutually exclusive, then probability of $A \cup B$ (i.e. either A or B occurring) is $P(A) + P(B) - P(A \cap B)$ where $A \cap B$ is the case where both A and B occur.

For example, if a coin is tossed twice (or two coins are tossed once each), we can define two events as: A, at least one head; B, at least one tail.

The probability of A occurring is $P(A) = \frac{3}{4}$
The probability of B occurring is $P(B) = \frac{3}{4}$
The probability of both A and B occurring is $P(A \cap B) = \frac{1}{2}$

Hence the probability of either A or B occurring is

$$P(A \cup B) = P(A) + P(B) - P(A \cap B) = \frac{3}{4} + \frac{3}{4} - \frac{1}{2} = 1$$

Additive model.   A model in which a number of factors (or variables) are assumed to have an additive effect on the dependent variable. For example, the usual linear regression model is additive, that is to say, the dependent variable is influenced by an aggregate of factors, indicated by adding together values of the regressors:

$$Y = \beta_0 + \beta_1 x_1 + \beta_2 x_2 + \beta_3 x_3 + \ldots + \varepsilon \tag{1}$$

This may be contrasted with, for example, the multiplicative model:

$$Y = \beta_0 . x_1^{\beta_1} . x_2^{\beta_2} . x_3^{\beta_3} . \ldots \; x_k^{\beta.k} . \varepsilon \tag{2}$$

The additive model, equation (1), may be derived from model (2) above by taking logarithms

$$\log Y = \log \beta_0 + \beta_1 \log x_1 + \beta_2 \log x_2 + \ldots \beta_k \log x_k + \log \varepsilon \tag{2a}$$

Agreement, coefficient of. A coefficient which measures the extent of agreement between preferences (or rankings) of people about objects or standards of performance. The relevant formula is

$$u = \frac{8 \Sigma A}{m(m-1).n(n-1)} - 1$$

where ΣA is the number of agreements between pairs of observers, m is the number of observers and n the number of objects. Thus, for example, if there are three objects A, B and C and two judges whose ranks are respectively A, B, C and A, C, B there is only one pair of agreements (A, A) and the coefficient is therefore

$$u = \frac{8 \times 1}{2 \times 6} - 1$$

$$= -0.33$$

The advantage of this generalized coefficient over KENDALL'S TAU is that it can be used for any number of judges. Assume, for example, that four judges give the ranks:

There are now four pairs of agreements (A,A; A,A; C,C; B,B) and the coefficient of agreement is therefore

$$u = \frac{8 \times 4}{12 \times 6} - 1$$

$$= -0.55$$

In the event of absolute agreement, e.g. (A,A; B,B; C,C) among two judges, the coefficient is

$$u = \frac{8 \times 3}{2 \times 6} - 1$$

$$= 1.0$$

Aggregation. The summing of statistical data for expression as a total value. If statistical data are homogeneous units, e.g., identical products of a manufacturer, aggregation presents no problems. Hence one can speak of production of 10 million litres of a given brand of paint in a year.

The aggregation of published statistics sometimes involves the addition of heterogeneous statistical units, as, for example, in the production of price indices where quantities of relevant consumer goods are weighted, or book-issue totals of libraries where data of different categories of non-fiction and fiction books are added together and expressed as an aggregate, or the summation of marks in different subjects in an examination. The problem of heterogeneity limits the usefulness of aggregate figures.

Aided recall. Any method used to help avoid RECALL LOSS. For example, in a study of readership of newspapers and magazines, respondents may be provided with a list of publications and asked separately about each. The problem with this procedure, however, is that respondents may be tempted to overstate their readership because of a desire not to appear uneducated or poorly informed, simply because of the presence of the list.

Algorithm. A formula or formulation of a problem. The term is frequently used in computing and in linear programming. In computing it specifically means a series of procedures or steps for the solution of a given problem. In linear programming, for example, it takes the form of a model:

Maximize (or minimize) a given set of values,
subject to a series of constraints.

The optimum solution in this case is approached by means of a series of stepwise or iterative procedures. (♦ SIMPLEX METHOD.)

Alienation. The extent of *non-correlation* as distinguished from *negative correlation*. Alienation is perfect if there is zero correlation. If a pair of variates is either perfectly positively or negatively correlated, alienation does not exist.

Alienation, coefficient of. A measure of alienation, related to the PRODUCT-MOMENT CORRELATION coefficient (r).

$$k = \sqrt{(1 - r^2)}$$

If r is either $+1$ or -1, $k = 0$. If $r = 0$, $k = 1$. There are no negative measurements. Hence if $r = 0.8$, $r^2 = 0.64$, $k^2 = 0.36$ and $k = +0.6$, but if $r = -0.8$, the result is similar; $k = +0.6$. The coefficient is a useful measure of non-dependence and non-determination.

Allocation of a sample.   The way in which the total sample is distributed across parts of the population (e.g. strata). For example, proportionate allocation involves selecting sample elements from each part of the population in proportion to the number of elements in that part of the population. ◗ OPTIMAL ALLOCATION, PROPORTIONATE STRATIFIED SAMPLING.

Allowable defects.   The critical level (or number) of defects in a sample batch. The number cannot be exceeded without there being either complete rejection of the batch or full examination of all items in it. In QUALITY CONTROL two limits are set, an acceptance limit and a rejection limit. If the number of defective items falls between the limits, sampling is continued. If it falls above the rejection limit the batch must either be completely rejected or each of its items must be scrutinized individually. If it falls below acceptance limit the lot is completely accepted. (◗ ACCEPTANCE SAMPLING.)

Alternative hypothesis.   In hypothesis testing any hypothesis other than the hypothesis being tested may be called an alternative hypothesis. Usually the hypothesis being tested is called the NULL HYPOTHESIS and the test consists of comparing the null hypothesis with the alternative hypothesis in the light of the sample observations.

Antecedent variable.   A variable which is antecedent in time, but not necessarily an 'independent' variable (or regressor). The variation of share values may, for example, depend on expectations of future results rather than past processes.

Antimode.   The opposite of MODE, i.e., a variate value for which the frequency distribution has a minimum. Although the zero tails of a normal frequency distribution are theoretically antimodes, the term is not usually used in this case.

Arbitrary origin.   A convenient reference point, used for the simplification of calculation. For example, when calculating the mean of the values 300, 353, 384 and 361, it is convenient to use one of the middle observations (e.g. 353) as a reference point to avoid summing large values.

In this case, the differences between the observations and the arbitrary origin, 353, are summed and divided by the number of observations. Since every observation has been decreased by 353, this is then added to obtain the arithmetic mean.

Example

Sum the deviations $-53 + 0 + 31 + 8 = -14$

Divide by the number of observations $\dfrac{-14}{4} = -3.5$

Add the arbitrary origin $-3.5 + 353 = 349.5$
The mean is thus: 349.5.

When it is used for calculating the mean, the arbitrary origin is often called an 'assumed mean'. The term 'arbitrary origin' is much more extensive than 'assumed mean', for it may refer to convenient reference points other than those used for calculating the moments of a frequency distribution. For example, the term 'arbitrary origin' is more appropriate than 'assumed mean' when applied to convenient reference points used for conversion of the values of x and y to simplify the calculation of regression and correlation coefficients.

Area sampling. The basis of survey sampling in many parts of the world where comprehensive population lists are not readily available. The basic procedure is that the geographical area under investigation (which could be as large as a country or as small as a town) is divided into smaller sub-areas of which a sample is selected by some probability method. Within each of these selected sub-areas, further stages may be introduced in which still smaller sub-areas are selected. Finally, the elements within the selected sub-areas may be listed and either a complete enumeration is carried out or an element sample selected. This is in essence MULTISTAGE SAMPLING, where the sampling units are areas and the sampling frame consists of maps rather than lists of elements.

Area sampling is the basic sampling method used for sample surveys in the United States, but although not widely used in Britain it could be very useful in some situations, for example in selecting samples of small businesses (groceries, garages, etc.).

Arithmetic progression. A series (or sequence) of numbers ranked in rising or falling order such that the intervals between successive pairs of terms are constant. Because the FREQUENCY DISTRIBUTION of the progression is only 1 for each value of the variate and the interval between each successive pair of values is constant, the MEAN and MEDIAN are equal and the sum of this kind of progression can be obtained by multiplying the mean or median by the number of terms.

As the value of the median is $\frac{1}{2}(x_1 + x_n)$ and there are n terms in the sequence, the sum of the series is

$$\frac{1}{2}n(x_1 + x_n)$$

Example

The series 3, 6, 9, 12, 15, 18 is an arithmetic progression because the interval between each successive pair (3,6), (6,9), (9,12) is constant at $+3$. If the numbers are added in the usual way, the sum is 63.

Using the formula

$$\tfrac{1}{2} \times 6 \, (3 + 18) = 3 \times 21$$
$$= 63$$

Arithmetic scale.   This is the correct term to describe the usual scale adopted in drawing graphs and diagrams. The axes are marked at intervals proportional to the numbers being plotted although the scale will often differ between the horizontal and vertical axes, e.g.

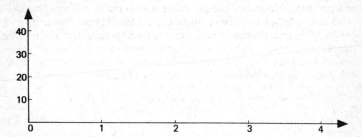

   Non-arithmetic scales such as the logarithmic scale do not have this equal spacing property.

Array.   A presentation of a set of values. For example, quantities of inventory items for the first six months of a given year are

   21, 7, 5, 19, 11, 20

In this cast, the array is a frequency array, the six categories being months 1 ... 6.
   Vectors and matrices are examples of arrays.

Association.   The term is often used in much the same way as CORRE-LATION but in a more particular sense it means the degree of inter-dependence or dependence between two sets of categories. The example in the entry CHI-SQUARED TEST (where schools are one set of categories and degrees of success in an examination are the other) illustrates this more particular meaning of the term, where there are more than two categories in each set.
   The simplest measures of association are concerned with dependence between two pairs of categories. These are explained in ASSOCI-ATION, COEFFICIENTS OF. (◆◆ CHI-SQUARED STATISTIC.)

Association, coefficients of.   Coefficients which measure the strength of ASSOCIATION between two sets of categories. The simplest measures of association are concerned with the strength of association between sets consisting of only two categories (e.g. presence or absence of an attribute).

Two possible coefficients for the simple case are given in the examples below.

Where either set contains more than two categories, more complex measures such as the CHI-SQUARED (χ^2) STATISTIC, which also applies to the 2 × 2 case, may be used.

Example

Let us consider two cases: (1) of thirty-two students who attempt two examinations (A and B), sixteen students are successful in both examinations, and sixteen are unsuccessful; and (2) of the thirty-two students, four are successful in both examinations (C and D), twelve are successful in examination C, twelve in examination D, and four are completely unsuccessful.

In case (1) there is perfect positive association and in case (2) there is negative association between the results of the two examinations.

Two simple coefficients M and Q may be used to measure association in both of the two cases. If we use the letters a, b, c and d to indicate the categories, in case (1) $a = d = 16$ and in case (2) $a = d = 4$, and $b = c = 12$.

$$\text{The first coefficient } M = a - \frac{(a + b)(a + c)}{a + b + c + d}$$

Thus in case (1) $M = 16 - 8 = 8$, the measure is positive and the results of the two examinations are positively associated; in case (2) $M = 4 - 8 = -4$ and we may conclude that there is negative association.

The disadvantage of M is that it is not very useful for comparative measurement.

$$\text{The second coefficient } Q = \frac{ad - bc}{ad + bc}$$

varies between −1 and +1.

Using this formula in case (1) of the above example:

$$Q = \frac{256}{256}$$
$$= +1$$

Association is perfect, a state resembling perfect positive correlation.

In case (2) of the example

$$Q = \frac{-128}{160}$$
$$= -0.8$$

It can be seen that the second coefficient Q is analogous to a correlation coefficient (i.e. values ranging from -1 to $+1$) and can easily be used for comparative purposes.

A further coefficient of association is discussed under COLLIGATION YULE'S COEFFICIENT OF COLLIGATION.

Assumed mean.   If numbers are so large that it is difficult to calculate the exact mean by the process of adding observed values and dividing by the number of observations, the procedure can be simplified by choosing (or estimating) an arbitrary origin (written X'), calculating the difference between each value and the arbitrary origin or assumed mean $(X_i - X')$, finding the mean of these differences and adding it to the assumed mean, such that

$$\overline{X} = X' + \Sigma \; \frac{(X_i - X')}{n}$$

Example
Calculate the mean of the following values:

	Normal method		Assumed mean method (use 60)	
	79			$+19$
	90			$+30$
	47		-13	
	59		-1	
	53		-7	
	62			$+2$
$\Sigma X_i =$	390	$\Sigma(X_i - X')$	-21	$+51$	$= 30$
$\dfrac{\Sigma X_i}{n} =$	65	$\dfrac{\Sigma(X_i - X')}{n}$			$= 5$

As X' was chosen to be 60 (see above), the exact mean is $60 + 5 = 65$.

The use of an assumed mean or abitrary origin is particularly useful where large numbers fall in a narrow range, e.g. 1350 to 1500.

In calculating variances, an assumed mean may also be useful since the variance of the original observations equals the variance of their differences from any assumed mean. Thus, without directly calculating the mean of the values given above, we can write

      Variance of $\{79, 90, 47, \ldots\}$
   = Variance of $\{19, 30, -13, \ldots\}$

Asymmetric.   This term is often employed as a synonym for 'skewed'. It is, however, more extensive in meaning than skewed. Any distribution is asymmetric where there is no central point of reference m, such that the frequency of the value $m - d$ is equal to the frequency of the value $m + d$.

Skewness is a particular form of asymmetry where there is only one mode; for example, where the shape of the frequency curve approximates that for a given binomial distribution where p is not equal to q.

Attenuation. The degree of underestimation or misestimation of correlation of bivariate data because of ERROR OF OBSERVATION.

Attribute. Attributes are qualitative or functional characteristics of individuals, objects or groups, as distinguished from quantifiable characteristics. Thus, for example, age, height, weight and wealth of individuals can all be regarded as variables because they can be quantified, but sex (i.e. male or female), origin (e.g. British, or West Indian, etc.), function (smoker or non-smoker), political persuasion (e.g. Labour or Conservative) can be regarded as attributes.

Attributes may be dichotomous or polytomous. Examples of dichotomous attribute pairs are: male and female; retired or non-retired; cancer patients or non-cancer patients. Examples of multiple or polytomous classifications of attributes are: blue-eyed, grey-eyed or brown-eyed; professional, manual, clerical, etc.; arts graduates, science graduates, law graduates, etc.

Autocorrelation. In data recorded sequentially through time (time series) the correlation between observations separated in time by a given amount or distance is called the autocorrelation of the series at that distance (lag). Sometimes the term is used more specifically to refer to the correlation between adjacent observations of a time series (lag 1). The autocorrelation of the residuals of a regression equation fitted to time series data is often estimated to provide an indicator of the goodness of fit; it being usually assumed that if the model is an adequate representation of the data, the autocorrelation of the residuals will be approximately zero. The collection of autocorrelation values at successive lags makes up the autocorrelation function. The autocorrelation between observations s time periods apart of a time series X_1, X_2, X_3, ... X_n may be estimated by

$$\rho(s) = \frac{\sum_{t=1}^{n-s} (X_t - \frac{1}{n-s}\sum_{j=1}^{n-s} X_j)(X_{t+s} - \frac{1}{n-s}\sum_{j=s+1}^{n} X_j)}{\sqrt{\left\{[\sum_{t=1}^{n-s}(X_t - \frac{1}{n-s}\sum_{j=1}^{n-s} X_j)^2][\sum_{t=s+1}^{n}(X_t - \frac{1}{n-s}\sum_{j=s+1}^{n} X_j)^2]\right\}}}$$

(◗◗ DURBIN-WATSON STATISTIC.)

Autoregression. A model that is sometimes used to represent a time series assumes that the value of the variable at time t is a proportion of its value in one or more previous periods plus a disturbance term representing the innovation or new factors affecting the variable, in the current period. A first-order autoregression relates the current observation to the immediately preceding one only, i.e.

$$X_t = \alpha \, X_{t-1} + u_t$$

where u_t is the disturbance term. Higher order models involve more lagged values on the right-hand side.

Average. A value which represents in typical (or summary) form the relevant characteristics of a set of values. In non-statistical usage (e.g. by accountants) the term usually means the MEAN, but in statistical usage it often means any measure of central tendency, such as the MEDIAN or MODE.

Average article run length. The average number of items sampled before action is taken in QUALITY CONTROL. The term should not be confused with AVERAGE RUN LENGTH.

Average deviation. This is preferably known as the MEAN DEVIATION.

Average linkage. A method of CLUSTER ANALYSIS in which the distance or similarity between an element and a cluster is calculated as the average distance or similarity between the element and all the elements which constitute the cluster. Thus in the diagram below, the distances between the element a and the elements in the cluster A (b, c, d) are given on the lines joining a, b, c and d. The distance of a from A is defined in this case as

$$1/3(ab + ac + ad) = 1/3[3 + 5 + 7]$$
$$= 5$$

The method is in essence a compromise between SINGLE LINKAGE, where the distance between an element a and a cluster A is defined as the shortest of the distances between a and any of the elements in A (in this case $ab = 3$), and COMPLETE LINKAGE, where the distance is defined as the longest of these distances (in this case $ad = 7$). Thus the clusters will be neither as compact as those formed by complete linkage nor as dispersed as those formed by single linkage but may provide an acceptable practical compromise between the two extremes.

Average Outgoing Quality Limit (AOQL).   The maximum proportion (or percentage) of defective items which can be accepted in a sampling inspection plan, i.e. the upper tolerance limit. It is equivalent to the maximum proportion (or percentage) of defective items which represent an average of outgoing quality.

Average point prevalence rate.   The simple average of the POINT PRE-VALENCE RATES for selected points within the observational period. The selected points will normally be, for instance, one day in each of the twelve months of a year or one day in each of thirteen weeks of a quarter.

Average Run Length (ARL).   At each given level of quality there is an average run length for a particular sampling inspection scheme. This is defined as the average (mean) number of samples of a given (constant) number of items which may be taken from the point at which the process begins to run to the point at which the sampling inspection scheme shows that the process is likely to have changed from acceptable to rejectable quality.

B

Balanced sample. A sample which has some predefined characteristics in common with the population from which it is drawn. If, for example, the mean age of the population is known, and the age of each sample member can be obtained, the sample is selected so that the sample mean age is equal (or nearly equal) to the population mean age. More than one variable can be used in the balancing procedure. In the case above, the proportion of males in the sample could also be made to correspond to the proportion of males in the population.

This method enjoyed considerable popularity at one time but its usefulness and validity may be questioned. It is not a PROBABILITY SAMPLING method and the fact that the sample and the population correspond in terms of some known characteristics provides no guarantee that the sample will give an accurate estimate of the variable of interest in the survey. In measuring attitudes, for instance, the variables which it is possible to use in selecting the sample may not be at all related to the attitudes being measured. Probability sampling with POST-STRATIFICATION is preferable. (◆ QUOTA SAMPLING.)

Band chart (or band curve chart). A special type of chart used for comparing and summing the values or measures of a set of constituents (or components) either over a period of time or (less usually) some other variate. The constituents of the set are plotted above each other so that the chart consists of a set of shaded, hatched or coloured bands. They may be used:

1. to show how absolute values of each component have changed over a period;

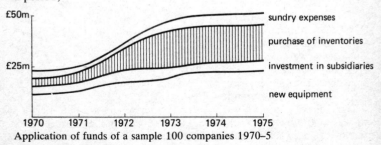

Application of funds of a sample 100 companies 1970–5

2. to show how relative values of each component have changed over a period.

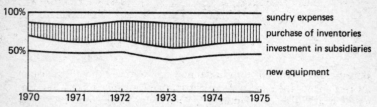

Data of sample 100 companies converted into percentage form

Bar chart. Similar to a HISTOGRAM, a bar chart is a pictorial representation of values of frequency measurements. The main difference between a bar chart and a histogram is that a bar chart exhibits each class in a distinct and disjoint form, whereas a histogram joins the class rectangles and uses a horizontal axis.

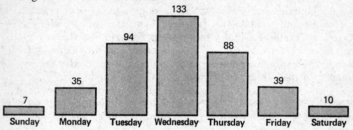

Example
In the above illustration 406 volunteers commence at 12 noon on a particular day to walk 200 miles to raise funds for a charity. The rectangles enable us to compare visually the numbers of volunteers who arrive at the destination in daily batches or classes. The advantage of a bar chart is that the eye can distinguish each group clearly, but as it has no axes it is preferable to show the frequency measures over or in each of the bars.

The bars may be horizontal or vertical, and the classes (or groups) may be structured by using intervals of time (as in the above case) or any other measurement.
Particular variations are:
1. a COMPONENT BAR CHART; and
2. a chart showing relative performance when measured against a mean, mode or other 'standard'.
The following is an illustration of (2) where the absolute performances of groups A, B, C and D are 200, 150, 70 and 220 respectively and the standard is the mean of 160, all four groups being equal in size.

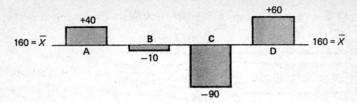

Because bar charts have no horizontal axes and can therefore display measurements for each class in distinct and disjoint form, they can, unlike histograms, be adapted to display measurements of ATTRIBUTE classes, as well as those of variate groups. For example, the following bar chart displays the workforce of a firm in four attribute classes (a) British-born male, (b) British-born female, (c) overseas-born male and (d) overseas-born female.

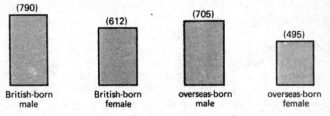

Base. A number used as an origin (or term of reference) for relative measurements of other values of a variate, or for the compilation of an index.

Example
The prices of a commodity may have been:

Year	£
1971	49
1972	50
1973	54
1974	60
1975	63

In this case, the 1972 price could be used as a base for the calculation of index numbers for future percentages, hence

Year	Index No. (1972 = 100)
1971	98
1972	100
1973	108
1974	120
1975	126

It is thus possible to see percentage increase between 1972 and each other year (8%, 20%, 26%, etc.). To convert values into index values the base becomes the denominator and the variate value to be converted becomes the numerator. Thus, for example, for 1975

$$\frac{63}{50} \times 100 = 126$$

Base line.   The essential horizontal line on any graph (or two-dimensional exhibit). It corresponds to the basic measurement of the variable (or variate) depicted on the vertical scale and is most often, though not necessarily, at the base of the graph. Where the vertical scale is an arithmetic scale the origin is usually understood to mean zero. The use of other measurements as origins can sometimes lead to exaggerated differences and misrepresented information. Thus the use of other origins for the base lines of graphs should either be avoided or, if used, should be indicated plainly.

Base period.   A reference period to which other periods are compared, often as a percentage with the base period equal to 100. In the construction of Laspeyres index numbers the base period is also used to obtain the weights which are then applied to the changes in the components of the index. Base periods are sometimes chosen to be years when data are particularly abundant, e.g. to coincide with a census of production or population.

Base weight.   A weight used in a composite aggregative index, calculated from the quantity information relevant to a base period. (◗ BASE PERIOD, LASPEYRES INDEX, PAASCHE INDEX, QUANTITY WEIGHTS.)

Batch variation.   The variation which occurs between products made in batches. Because the products are made in batches, and not either singly or continuously, batch variation is a composite of two kinds of variation:
1. intra-batch (or intra-class) variation – that is, the variation which occurs between the units (or items) within each batch; this may be accidental or may reflect inherent variation in the manufacturing process;
2. inter-batch (or inter-class) variation – the variation occurring between one batch and another; although this could also be random or reflect variations in the manufacturing process, it may also be the result of differences in the quality of material input.

Bayes' postulate.   Bayes (1702–61) suggested that the prior probabilities of a set of mutually exclusive propositions should be assumed equal where nothing is known to the contrary. This assumption is known as Bayes' postulate and is crucial in the Bayesian approach to inference. (◗ BAYES' THEOREM.)

Bayes' theorem. Suppose we have an exhaustive and mutually exclusive set of events B_1, ... B_r, ... B_n. Each of these has associated with it a probability $p(B_r)$. Also, the event A is possible given any of the events B_1, ... B_n but the probability of A occurring is different for different events B_r ($r = 1$, ... n). Next, suppose that the event A has actually occurred. The theorem enables us to calculate the probabilities of the various B events. It states the probabilities of each of the events B_r, given that the event A has happened.

The set of probabilities $\{P(B_r)\}$ is known as the set of *prior probabilities*; the probabilities $\{P(B_r|A)\}$ are the *posterior probabilities*, i.e. posterior to A having occurred. The probability $P(A|B_r)$ is the likelihood. Thus Bayes' theorem states that the posterior probability is proportional to the prior probability multiplied by the likelihood.

If we have a set of mutually exclusive events B_1, B_2, ... B_n and another event A, the multiplicative rule of probability tells us that

$$P(B_r,A) = P(A)P(B_r|A) \tag{1}$$

and

$$P(B_r,A) = P(B_r)P(A|B_r) \tag{2}$$

Thus

$$P(B_r|A) = \frac{P(B_r)P(A|B_r)}{P(A)} \tag{3}$$

If the events B_1, ... B_r, ... B_n are exhaustive, then summing equation (3) over the B events, we have

$$\sum_r P(B_r|A) = 1$$

$$= \sum_r \frac{P(B_r)P(A|B_r)}{P(A)}$$

Hence

$$P(A) = \sum_r P(B_r)P(A|B_r) \tag{4}$$

Substituting equation (4) in equation (3), we have

$$P(B_r|A) = \frac{P(B_r)P(A|B_r)}{\sum_r P(B_r)P(A|B_r)} \tag{5}$$

which from equation (2) gives

$$P(B_r|A) = \frac{P(B_r,A)}{\sum_r P(B_r,A)} \tag{6}$$

Bayes' theorem is expressed in the equations (5) and (6). It states that the probability (P) of B_r, given A, is proportional to the probability of B_r multiplied by the probability of A, given B_r.

The difficulty with Bayes' theorem is not with its mathematical formulation but with some of its applications. The theorem assumes that the prior probabilities $P(B_r)$ are known, as they must be if we are to calculate the posterior probabilities. However, in general they are not known and thus to use them we must assume something about them. Bayes' suggestion, embodied in BAYES' POSTULATE, was that unless something is known to the contrary, we should assume that they are equal. It is this 'Principle of the Equidistribution of Ignorance' which has led to the considerable controversy surrounding the Bayesian approach to statistical inference.

Bell-shaped curve. Symmetrical continuous frequency distribution resembling the diagram of a vertical cross-section through a bell.

Bernoulli's theorem. The theoretical basis for estimating parameters of the binomial probability distribution. The theorem states, in effect, that if P is the proportion of a population with a given attribute and if $Q = 1 - P$, then the probability that the sample proportion (p) possessing the attribute is close to P tends to 1 as the sample size (n) tends to infinity. This is because the variance of the sampling distribution of p is $\dfrac{PQ}{n}$, which tends to zero as n tends to infinity.

Bessel's correction. Introduced into measures of dispersion when calculated from a sample in cases where the population mean μ is not known. Since part of the information from the sample is used in estimating μ the population mean, using $\overline{X}$ the sample mean, there are only $n - 1$ independent pieces of information left to estimate dispersion. Thus it is necessary to use the correction $\dfrac{n}{(n - 1)}$ in calculating these measures.

The variance of the values in a particular sample is multiplied by $\dfrac{n}{(n-1)}$ to provide an unbiased estimate of the variance in the population from which the sample was drawn.

Therefore, for example, the variance estimated from a given sample:

$$s^2 = \frac{\Sigma(X_i - \overline{X})^2}{n} \times \frac{n}{n - 1}$$

$$= \frac{\Sigma(X_i - \overline{X})^2}{n - 1}$$

(➤➤ DEGREES OF FREEDOM.)

Beta-coefficient. A dependent variable (Y) is often expressed as the sum of the products of independent variable (regressor) values and their coefficients, and a regression constant ($\blacklozenge$ 'a' as a REGRESSION CONSTANT). In regression equations the Greek letter β (beta) is often employed to denote the coefficient of a variable, that is, the quantity by which the particular value of a variable must be multiplied.

A multiple linear regression equation may thus be written:

$$Y = \beta_0 + \beta_1 x_1 + \beta_2 x_2 + \beta_3 x_3 + \ldots \beta_p x_p + \varepsilon$$

where Y represents the dependent variable; x_j represents a given independent variate ($j = 1 \ldots p$); β_0 represents the regression constant; β_j represents each coefficient for a given independent variate; and ε represents the error (or disturbance) factor.

The simplest case where beta-coefficients are used is in a bivariate regression model where $Y = \beta_0 + \beta_1 x$. The beta-coefficient of x, ($= \beta_1$) may be estimated from a sample, using the formula

$$\beta_1 = \frac{n\Sigma xY - (\Sigma x \Sigma Y)}{n\Sigma x^2 - (\Sigma x)^2}$$

where Y is linearly CORRELATED with x and dependent on it.

Example

Observation	x	Y	x^2	Y^2	xY
1	8	17	64	289	136
2	10	20	100	400	200
3	4	11	16	121	44
4	6	14	36	196	84
	$\Sigma x = 28$	$\Sigma Y = 62$	$\Sigma x^2 = 216$	$\Sigma Y^2 = 1006$	$\Sigma xY = 464$

$$\beta_1 = \frac{(4 \times 464) - (28 \times 62)}{(4 \times 216) - (28)^2}$$

$$= \frac{(1856 - 1736)}{(864 - 784)}$$

$$= 1.5$$

In this case, β_0 can be estimated as $\overline{Y} - (\text{est } \beta) . \overline{X} = 5.0$.

Beta-error (or β-error). An error which results from accepting the NULL HYPOTHESIS when it should be rejected. Beta errors are also known as errors of the second kind, or as type II errors. ($\blacklozenge$ HYPOTHESIS TESTING.)

Between groups sum of squares.　In the analysis of variance of data consisting of a number of groups the sum of squares about the mean of the observations from all groups is the aggregate of:

1. the sum of squares of observations about group means; and
2. the sum of squares of group means about the mean of all data.

Category (1) divided by the appropriate number of degrees of freedom is variously termed:

(a) within-groups mean-square;
(b) internal mean-square; or
(c) intraclass mean-square.

Category (2) divided by the appropriate number of degrees of freedom is alternatively termed:

(a) between-groups mean-square;
(b) external mean-square; or
(c) interclass mean-square.

The term VARIANCE is sometimes used in place of mean-square in the above expressions.

Bias.　A systematic and non-random (but not necessarily intentional) distortion in a result or sample. Statistical bias must be distinguished from non-statistical bias. In either case, bias means that the mean of a large number of sample estimates does not approach the population parameter being estimated.

Statistical biases arise when the expectation of the sampling estimate does not equal the population parameter being estimated, when, for example:

1. no allowance is made for non-proportional allocation in a stratified sample;
2. the sample mean is divided by the number of observations to obtain the population variance instead of by one less than the number of observations $(n - 1)$. (◗ BESSEL'S CORRECTION.)

The first source of bias (1) does not decrease to zero as sample size increases so the estimator is asymptotically biased, whereas the second source of bias (2) reduces with larger samples so the error resulting from the estimate renders it asymptotically unbiased.

Non-statistical biases result from such causes as biased questions, defective sampling frames and non-response. In general, information from some source external to the investigation itself is necessary if we wish to estimate non-statistical biases.

Biased sample.　A sample selected using a BIASED SAMPLING METHOD. The term is somewhat unsatisfactory since it is the sampling method which is biased rather than the sample itself.

Biased sampling method. A sampling procedure which systematically discriminates in a non-quantifiable way against some part of the population. Consequently samples obtained using the sampling procedure will be unrepresentative of the population.

We should distinguish between sampling with known unequal probabilities (e.g. disproportionate allocation of a stratified sample or sampling with probability proportional to size) and biased sampling. In the former situation, the over-representation of certain parts of the population can be allowed for in the estimation procedures; in the latter, the error caused by the bias (which is non-measurable) is irretrievable.

QUOTA SAMPLING, for instance, is an example of biased sampling, since the probabilities of selection of the population elements are not known and the sample is subject to selection bias.

Bimodal. Having two modes. For example, a distribution curve may consist of two modes when it is effectively the composite of two distribution curves. The modal weights of males and females of a given height and age are different. A composite sample distribution of the weights of all persons (i.e. male *and* female) of a given height and age may therefore be bimodal. (◗ MODE.)

Binary. Any system is binary when its constituents are primarily and sequentially divisible by using two, and only two, categories.

Examples are:
1. the binary notation;
2. binary sequences; and
3. binary experiments.

The binary notation is the basis of computer calculation. Computers operate by using only two 'switch-circuit' states, 'open' and 'closed', and there is therefore a need to express all numbers in a form which uses only two digit-characters, i.e. 0 and 1. Each power of 2 therefore involves the use of an extra digit;

decimal number	1	2	3	4	5	6	7	8	9	10	16	32
binary equivalent	1	10	11	100	101	110	111	1000	1001	1010	10000	100000

Numbers are *added* by simple application of the process $1 + 1 = 10$. Thus $5 + 7 = 12$ is calculated.

Decimal	Binary
5	101
7	111
12	1100 = (1000 + 100)
	8 + 4 = 12

Numbers are *multiplied* simply by applying the principles:

$$1 \times 1 = 1; \quad 1 \times 10 = 10; \quad \text{and } 1 \times 0 = 0$$

Binary sequences are those where each of the numbers forming the sequence can take only one of two possible values (e.g. 0 or 1). In a probabilistic sense, the outcomes of casting a coin successively (for example, HTTHHHTHHT), where H represents the outcome of a 'head' and T a 'tail' could be described as a binary sequence.

Finally, binary experiments are those where the outcomes of the experiment fall into one of only two categories.

Binomial.   Consisting of two terms. As a substantive the word is used to mean any algebraic expression consisting of two terms joined by a + or −. Thus $(x + y)$, $(m + x)$, $(a - b)$ are all binomials, or binomial expressions. The terms need not be equal in any respect; the expansion of two equal terms $H + T$, where heads and tails are involved, is a specific case of a binomial expression which, as the power to which it is raised approaches infinity tends to the normal distribution. (◗ BINO-MIAL DISTRIBUTION, BINOMIAL TEST and PASCAL TRIANGLE.)

Binomial distribution.   A special type of probability distribution to be applied to an important type of experiment consisting of a number (or sequence) of simpler experiments. In each of these simple experiments there are two possible outcomes which are the same for the whole sequence. The probabilities of the two outcomes, conventionally called 'success' and 'failure' remain the same throughout the sequence, and are usually denoted by p and $(1 - p)$ respectively. The binomial distribution gives the probability that in the set of say n trials, there will be r successes, where r can take on any value between 0 and n.

For example:
1. In tossing a fair coin we could label a head outcome, 'success' with $p = \frac{1}{2}$. But suppose we tossed it a number of times, say three, then the binomial distribution gives us the probabilities of observing 0, 1, 2, or 3 heads in this sequence.
2. The probability of a defective item in an assembly line could be $\frac{1}{8}$. What is the probability of, say, ten defective items in a sampled batch of 20?

In the coin tossing example we could list the possible outcomes of the sequence of trials {H,H,H} {H,H,T} {H,T,H} {T,H,H} {T,T,H} {T,H,T} {H,T,T} {T,T,T}. Using the rules of probability each outcome has a probability of $\frac{1}{2} \times \frac{1}{2} \times \frac{1}{2} = \frac{1}{8}$. The first corresponds to three heads, the last to no heads, where there are three outcomes for one head and two heads, corresponding to the outcomes of the simpler experiments in the sequence of three. The full probability distribution is then P (0 heads) = $\frac{1}{8}$, P (1 head) = $\frac{3}{8}$, P (2 heads) = $\frac{3}{8}$, P (3 heads) = $\frac{1}{8}$.

This is a special example which can be generalized for any value of p and for any number of trials n. With the coin tossing example in mind, suppose that the coin was not fair and did not fix p at $\frac{1}{2}$. From the rules of probability the outcome above would have probabilities p^3, $3p^2(1 - p)$, $3p(1 - p)^2$ and $(1 - p)^3$ respectively. The multiplier 3 for 1 head and 2 heads corresponds to the number of rearrangements of 1 head and 2 tails, and 2 heads and one tail respectively. In general with n trials the probability of r successes would be $p^r (1 - p)^{n-r}$ multiplied by the Number of rearrangements or orderings of r successes and $n - r$ failures. A formula can be developed for this from what are known as binomial coefficients, and is

$$\frac{n!}{r!(n - r)!}$$

$n!$, $r!$, $(n - r)!$ are to read as n factorial, r factorial, $(n - r)$ factorial respectively. Examples of factorials are $6! = 6 \times 5 \times 4 \times 3 \times 2 \times 1$, $4! = 4 \times 3 \times 2 \times 1$ and so on.

The general formula for the binomial distribution is then:

$$\text{Probability of } r \text{ successes} = \frac{n!}{r!(n - r)!}p^r(1 - p)^{n-r}$$

In the assembly line example $n = 20$, $p = \frac{1}{8}$ and

$$\text{probability of 10 successes} = \frac{20!}{10! \ 10!}(1/8)^{10}(7/8)^{10}$$

(◆◆ COMBINATORIAL.)

Binomial test.   A test of a sample proportion against a hypothesized population proportion involving the use of binomial tables. The test is usually used where the hypothesized population proportions are not less than 10 per cent or greater than 90 per cent. Where they exceed these limits (for example 3 per cent and 97 per cent), tables based on the POISSON DISTRIBUTION are more appropriate.

A further limitation of the use of the test is that when the expected sample proportion exceeds the actual sample proportion, the number of items in the smaller sample category should not exceed 20, and when the actual sample proportion exceeds that expected (i.e. calculated from the hypothesized population proportion) the number of items in the smaller sample category should not exceed 4. If the number of items exceeds these limits, the Z-SCORE STATISTIC should be used.

Tables of critical values are usually divided into two parts, showing the highest expected number in a sample (n) at 5 per cent and 1 per cent significance levels, given a number in the smaller sample category (r), (1) when it is less than the expected number ($r<e$) and (2) when it is greater than the expected number ($e<r$), where e is the expected number, calculated from the hypothesized population proportion.

Example

Suppose that 20 per cent of students in a given area are known to fail a particular professional examination. Does the occurrence of only two failures in fifty candidates from Brightests' Supercollege represent a significant difference?

The expected smaller category (e) in this case is 20 per cent of 50 ($=$ 10). This value is greater than that observed ($r = 2$) so that $r<e$. As r ($= 2$) does not exceed 20 we may apply the test and use the section of the appropriate statistical table giving values when $r<e$. These show the upper critical values of n (the number of elements in the sample) at 5 per cent and 1 per cent significance levels to be 24 and 33 respectively. The actual sample size ($n = 50$) is greater than 33. The results are such that, under the NULL HYPOTHESIS, if Supercollege students have the same failure rate as other students, a 1/100 fluke is required to account for them; the null hypothesis is thus rejected.

Biserial correlation. The correlation of two or more dichotomous attribute measures. For example, sex (male and female) may be correlated with performance (success or failure) in a given examination.

Sometimes in the early literature the term was employed to mean the correlation of a variate with an attribute, as, for example, correlation between sex and grades of performance in a particular examination.

An advanced formula may be used to estimate the equivalent of the PRODUCT MOMENT CORRELATION coefficient where bivariate information is limited to a 2×2 table. This is sometimes known as tetrachoric correlation. For ordinary elementary statistical purposes the computation of a correlation coefficient would not be necessary because chi-squared and other tests could be used to test for significant association between category and attribute groupings.

Bivariate normal distribution. A two-way extension of the NORMAL DISTRIBUTION in which:
1. the MARGINAL DISTRIBUTIONS are normal;
2. the CONDITIONAL DISTRIBUTIONS are normal.

If the CORRELATION COEFFICIENT (ρ) between the two variates is zero, the two variates are independent of each other and the bivariate distribution is the product of two independent normal distributions.

It is the only case where zero correlation implies INDEPENDENCE.

Block. A group of (possibly adjacent or continuous) items in a given population or experimental set. In area sampling, a group of adjacent houses may constitute a block. In audit sampling, a series of vouchers (e.g. 2901 to 2950) may be subjected to detailed examination, a process known as 'block-sampling'. or 'block-vouching'. In medical investigation a population may be divided into blocks on the basis of hereditary or other factors.

The purpose of blocks in both sampling and experiments is to isolate a source (or sources) of variation. Constituent units in a block are considered homogeneous for the purpose of investigation to discover presence or absence of a particular factor, and particular forms of analysis (such as ANALYSIS OF VARIANCE) are employed to isolate inter-block variation (i.e. variation between blocks) and intra-block variations (i.e. variation within blocks). Inter-block variation will reveal particular blocks which can be examined more intensively.

Block diagram. A number of rectangles of differing vertical sizes joined to each other and placed upon a common base line. The important difference between a block diagram and a bar chart is that the rectangles of a block diagram are joined to each other, whereas those of a bar chart are not joined to each other.

Block diagrams may be used to display frequency or other measurements of attribute classes. Such would, for example, be the case if the four attribute classes (British-born male, etc.) in the relevant bar chart exhibited under BAR CHART were joined, rather than disjoint. Even when the horizontal axis is used to measure interval classes of a variate, the height measurements of each rectangle need not represent frequency measurements. For example, the horizontal axis may be used to measure age-interval groups (e.g. 20–29, 30–39, 40–49) and the heights of the rectangles would be used to depict the mean blood-pressure measurements for each age group.

A HISTOGRAM (with equal class intervals) is a particular form of block diagram where the horizontal axis is used to measure a variate in interval groups, and the heights of the rectangles are used to display frequency measurements.

BLUE – Best Linear Unbiased Estimator. In comparing different estimators of a population parameter we can restrict the estimators being considered to the class of LINEAR ESTIMATORS. We can further restrict the comparison to those estimators which are unbiased. The estimator which has the minimum variance of all the estimators in this class of linear unbiased estimators is called the 'best linear unbiased estimator', or BLUE. The LEAST-SQUARES ESTIMATOR of the population regression coefficient, is the BLUE. (◆◆ GAUSS-MARKOV THEOREM.)

Bounded recall.   A method for dealing with the problem of TELE-SCOPING in memory questions. There is a tendency for respondents to include in their answers to questions such as 'How many times did you go to the theatre in the last month?' events that occurred before the reference period. Bounded recall requires more than one interview and is therefore generally used only in PANEL STUDIES when each respondent is approached more than once. At the first interview, the question is asked and the answer recorded. At the second interview, say one month later, the question is asked again and the respondent is reminded of the answer given on the previous occasion. Thus duplication of information can be checked and the method will reduce substantially and possibly eliminate telescoping errors from the second response.

Bracket codes.   A CODING METHOD in which a number of (neighbouring) values are grouped together as a single code. The coding of age into age groups or income classes are examples. Thus the ages 0–9 might be coded as 1, 10–19 as 2, etc. and the full age distribution might be coded as:

Age	Code
0–9	1
10–19	2
20–29	3
30–39	4
40–49	5
50–59	6
60 and over	7

The use of bracket codes leads to a more efficient storage of information when the fully detailed information (specific values) is not required in the analysis.

Branching process.   A process by which frequencies in an expected future population are generated by probable outcomes or future states of members of an existing population, such that each category of members of the existing population has a number of expected states or outcomes.

For example, of a given set of customers at time 0, 0.5 will pay their accounts by time 1, the remaining 0.5 will wait until time 2, when 0.9 will pay and 0.1 will prove insolvent. Of those who pay their accounts at time n, 0.8 will reorder and constitute customers in time $n + 1$.

The process may be illustrated as follows:

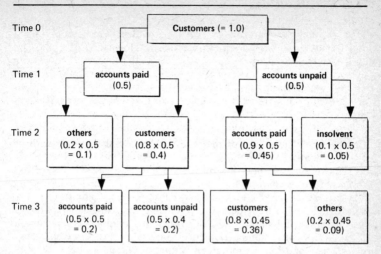

Thus at time 3 the distribution of the particular account-customers which existed at time 0 becomes:

New accounts paid in time 3	0.20
New accounts unpaid in time 3	0.20
New business in time 3	0.36
Insolvent original accounts	0.05
Others (0.1 in time 2 + 0.09 in time 3)	0.19
Total	1.00

Bulk sampling. The sampling of materials whose population is measured in bulk rather than in discrete units. For example, whereas cars, television sets, books, houses, etc. may be measured in units, other manufactures such as oil, whisky, wool, corn, etc. would be measured in quantities rather than in units. Thus a sample of a consignment of crude oil may be tested for its kerosene content, a shipment of whisky for its alcohol content or a sample may be taken from a domestic reservoir of water and tested for its fluoride content.

C

Call-backs. The failure of the investigator to make contact with all the designated respondents at the first attempt results in NON-RESPONSE. One method of reducing the non-response is to make one or more further attempts to contact the designated individuals or units. These later attempts are termed call-backs. They are an essential feature of all well-conducted surveys. On the first call, many potential respondents will be absent, busy or otherwise temporarily unavailable for interview. It is standard practice to make at least three calls (two call-backs), and four or more calls would not be unusual.

Cartogram. The depiction of statistical data of a descriptive nature on maps by means of symbols, dots, dashes or mass miniature drawings of the variable indicated. This method is particularly suited to branches of knowledge with a geographical content, e.g. economic geography, social geography, etc.

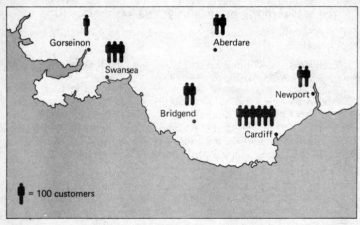

South Wales cartogram – indicating customers of Dexon Limited in a given year

Category. A convenient classification or grouping of a section of a population, usually for the purpose of frequency measurement.

Categories may either be attribute (e.g. blue-eyed, brown-eyed or grey-eyed; or male or female) or they may be interval measurements (e.g. salary categories using the groupings £0–£1999; £2000–£3999; £4000–£5999; and so on).

Cauchy distribution. A particular case of the STUDENTS t DISTRIBUTION with one degree of freedom. It has no finite moments apart from the mean.

Causal model. A MODEL in which the relationships involved are assumed to be causal – in other words, unidirectional relationships where a change in one variable causes the change in another. The causal structure must be specified in advance of the analysis – in cross-sectional data statistical methods do not permit the determination of CAUSAL RELATIONSHIPS but do allow the magnitudes of the effects to be estimated once the direction of causation has been specified or assumed. (♦ PATH ANALYSIS, PATH MODEL.)

Causal relationship. A relationship between two or more variables in which the direction of causation is specified. A change in one variable is assumed to lead to (or cause) a change in the other. For example, in specifying the relationship between father's occupation and son's occupation we might assume that it is the former which influences the latter and not vice versa. The specification of causation is outside the scope of statistics and lies in the subject matter area being examined. Statistical procedures do enable us to estimate the magnitude or strength of the relationship once the model has been specified.

Cell. A sub-class or sub-category in a two-way or multiway frequency classification.

Cell frequency. The number of observations which fall in a particular cell of a frequency classification.

Censoring. The discarding of the values of some constituents of a population or sample. It may, for example, be decided to choose a particular value of a variate and use only those values which fall within a given absolute distance from this fixed value.

Census. A complete enumeration of a population at a point in time. The enumeration should refer to well-defined characteristics. The best-known census is the Census of Population, which is a complete enumeration of all persons living in a country on a particular date with respect to basic demographic data. Similarly we have the census of production, the census of distribution, etc. A census can be contrasted with a SAMPLE, in which only a part of the population is enumerated or observed.

However, the term census is sometimes used to denote the kind of information which is collected. Censuses in that sense are concerned

with simple or basic information – largely with questions of counting
the numbers of elements with particular characteristics. In this context
SAMPLE CENSUS becomes a reasonable term.

Centiles (or percentiles).   One of a class of terms known as QUANTILES
or partition values which divide the total frequency of a population (or
sample) of a variate into n equal proportions, in this case into 100
equal proportions, when arranged in rank order.

Example
A comprehensive school contains 999 students, ranked according to
success in a particular examination.

Number of students	Grade
5	A
15	B
300	C
650	D
20	E
9	F

Centiles would occur at each tenth unit, since $999 + 1 = 1000$. Hence
the first and second centiles would be at B grade, the third a C grade,
etc., and the final centile an E item. There would be no Fs (failures ?)
among the centiles.

Centile measurements are most commonly used in education.

Central limit theorem.   A mathematical result of great practical
importance in statistical inference since it enables one to assert that the
sampling distributions of many statistics are approximately normal. If
there is a sequence of n random variables with identical distributions,
then the distribution of their mean approaches that of the normal
distribution as the size of n increases. Further, the mean of this normal
distribution is equal to the mean of the distribution of the variables,
and if σ is the standard deviation of the original distribution, then $\sigma/\sqrt{n}$
is the standard deviation of the limiting normal distribution. Nothing
need be assumed about the shape or form of the original distribution of
the variables.

Example
Suppose that in a population of 1 million families the proportions with
0, 1, 2, 3, 4 children are 0.2, 0.4, 0.2, 0.1 and 0.1 respectively. The
mean of this distribution can be calculated as $\mu = 1.5$. The means of
samples of size 100 from this distribution will have approximately a
normal distribution. The mean and standard deviation of this normal
distribution will be 1.5 and $\sigma/\sqrt{n} = 1.204/10 = 0.1204$, respectively.
Results such as this enable the statistician to make inferences about

the parameters of populations by reference to known properties of the normal distribution.

Central tendency. The tendency of quantifiable (or measurable) data to be bunched around a central (or typical) value, often described as an 'average'. Measures of location or of central tendency include the MEAN, MEDIAN and MODE. The explanation of this tendency may be read in entries BELL-SHAPED CURVE, BINOMIAL DISTRIBUTION, PROBABILITY DISTRIBUTION, NORMAL CURVE.

Measures of proximity of variate values to central value (or conversely of non-proximity or dispersion) are MEAN DEVIATION, RANGE, STANDARD DEVIATION, VARIANCE and QUARTILE DEVIATION.

Centred average. A term used in TIME SERIES ANALYSIS. The analysis of time series involves the isolation of several factors, including (1) a trend line, (2) seasonal fluctuations, (3) cyclical fluctuations, (4) episodic movements and (5) random fluctuations.

The isolation of seasonal fluctuations could be effected by using moving averages of the four quarters of a year if it were not for the fact that there are four seasons (or twelve months) in a year and that division by an even number (four or twelve) does not give a representative figure for any particular quarter or month. It is thus necessary to use an average of moving averages.

Example

year	\multicolumn{4}{c} 1975/6				\multicolumn{4}{c} 1976/7				1977/8	
quarter	1	2	3	4	1	2	3	4	1	2
sales (units)	1176	990	800	1414	1224	1030	832	1470	1272	1070
moving average			1095	1107	1117	1125	1139	1151	1161	
centred average			1101	1112	1121	1132	1145	1156		
variation factor (%)			72	127	109	92	73	128		

The moving average of quarters 1 to 4 of 1975/76 cannot be compared with any of the values for sales in respect of the quarters themselves since it gives a theoretical value appropriate to the end of quarter 2 and the commencement of quarter 3. The trend figure for quarter 3 can only be obtained by calculating the average of that at the commencement of quarter 3 (i.e. 1095) and that at the beginning of quarter 4 (i.e. 1107). The resultant centred average (i.e. 1101) can be compared with the actual sales figure for that quarter (800) and the extent of variation calculated

$$\left(\frac{800}{1101} = 72\%\right).$$

Chain indices.   Given indices for the percentage change between consecutive pairs of time periods it is possible to multiply these together to form a chain index. For example, if a Laspeyres index of prices in 1977 based on 1976 = 100 was 107 and a similar index for 1978 based on 1977 was 110, then the chain index would be:

$$1976 \quad 1977 \quad 1978$$
$$100 \quad \ 107 \quad \ \ 117.7 \ (= 107 \times \frac{110}{100})$$

In this case we are essentially splicing together indices for successive pairs of years and the resulting chain index has the advantage that its quantity weights are constantly being updated. Because of this it does not fossilize into an out-of-date expenditure pattern as, for example, the Laspeyres index would do. However, it is more risky to make comparisons between widely separate time periods because it is increasingly difficult to be sure that only price movements are being compared rather than quantity and quality changes, these being a reflection of changing tastes.

Chaining effect.   Chaining effects may arise when the inclusion of an element in a cluster depends only on its relationship with only a single element belonging to that cluster. Thus the cluster formed may contain elements only distantly related to one another. The situation is represented diagrammatically below where the numbers allocated to the elements indicate their order of inclusion in the cluster. It can be seen that elements 1 and 8 are very far apart from one another, although both belong to the cluster. (◆ CLUSTER ANALYSIS.)

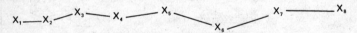

Charlier's check.   A useful check in the calculation of STANDARD DEVIATION and VARIANCE. It relies on the fact that

$$\Sigma f(x + 1)^2 = \Sigma fx^2 + 2\Sigma fx + \Sigma f$$

When calculating standard deviation using an ASSUMED MEAN or ARBITRARY ORIGIN it is important to be able to check the calculations of Σfx^2, Σfx and Σf with some speed. The simplest method of doing so is to construct a fourth frequency column $\Sigma f(x + 1)^2$.

Example
Assume that a variable has the following values and frequencies.

Value	Frequency
1	2
2	4
3	4

Although the mean is 2.2 we may wish to use the value 2 as an arbitrary origin and apply the formula

$$\sigma = \sqrt{\left[\frac{\Sigma fx^2}{\Sigma f} - \left(\frac{\Sigma fx}{\Sigma f} \right)^2 \right]}$$

Difference from origin (x)	Frequency (f)	fx	fx^2	$f(x + 1)^2$
-1	2	-2	2	0
0	4	0	0	4
1	4	4	4	16
	10	2	6	20

Applying Charlier's check

$$20 = 6 + (2 \times 2) + 10$$

In this case

$$\sigma = \sqrt{(0.6 - 0.04)}$$
$$= \sqrt{0.56}$$
$$= 0.7483$$

Chi-squared.   Strictly speaking Chi-squared, written as χ^2, is the sum of squares of independent standard normal variates. However, the term generally refers to a sum of the form

$$\chi^2 = \Sigma \frac{(O_{ij} - E_{ij})^2}{E_{ij}}$$

where $i = 1, \dots m$ and $j = 1, \dots n$ and where for a two-way classification O_{ij} is the observed frequency, and E_{ij} is the expected (theoretical) frequency under some hypothesis.

The χ^2 statistic is simply the value of χ^2 for any given data set. The formula given above refers to an $m \times n$ cross-classification, but the statistic can be calculated for any (mutually exclusive) classification. Indeed one of its most common applications is in comparing a grouped frequency table with some theoretical distribution. In this case the χ^2 statistic takes the form

$$\chi^2 = \sum_{j=1}^{k} \frac{(O_j - E_j)^2}{E_j}$$

where $j = 1, \dots k$ denote the categories in the frequency table, and O_j and E_j are the observed and expected frequencies respectively.

The χ^2 statistic can be used as a measure of association. In a two-way table the null hypothesis could be that of no association between the two classifications, the χ^2 statistic calculated and the χ^2 test applied. (♦♦ CHI-SQUARED TEST.)

Chi-squared distribution. The distribution of the sum of squares of v independent normal variates expressed in standard normal form, where v is the number of DEGREES OF FREEDOM. (♦ CHI-SQUARED TEST.)

Chi-squared test. This test is used for discontinuous or classified data consisting of mutually exclusive categories (or attributes). A researcher may, for example, be concerned with the results of a particular examination, and classify candidates by means of grade achieved, age (e.g. under 21 or over 21), race, sex, school or college, and may used the test to examine hypotheses whether males fail more often than females, or whether younger candidates have a better chance of distinction than older ones, etc.

Example

	actual examination results credit	pass	fail	total
Yellowfriars	25	148	27	200
Grimgrind	5	77	18	100
total	30	225	45	300

Procedure
To find out whether there is a significant difference between the schools' results:

1. Use the total column and row proportions to redistribute the results and thus calculate expected frequencies.

	expected examination results credit	pass	fail	total
Yellowfriars	20	150	30	200
Grimgrind	10	75	15	100
total	30	225	45	300

2. Calculate the differences between the corresponding elements of the two matrices (except the totals) square them and divide by the expected result. Then find the total, thus:

$$\chi^2 = \frac{5^2}{20} + \frac{-2^2}{150} + \frac{-3^2}{30} + \frac{-5^2}{10} + \frac{2^2}{75} + \frac{3^2}{15}$$

$$\chi^2 = \frac{25}{20} + \frac{4}{150} + \frac{9}{30} + \frac{25}{10} + \frac{4}{75} + \frac{9}{15}$$

$$= 4.73$$

3. Now consult χ^2 tables. DEGREES OF FREEDOM are $(c-1)(r-1) = 2 \times 1 = 2$. With two degrees of freedom at a significance level of 5 per cent the critical value of χ^2 is 5.99. As the observed value of χ^2 is 4.73 the results of Yellowfriars and Grimgrind schools are not significantly different at the 5 per cent level.

Note. If any of the expected values is less than 5, the test may not be valid.

Circular diagram/pie diagram. A pictorial method of comparing category frequencies with total value or total frequency, by showing each of the category frequencies as areas between the segments and the centre of a circle or as sections of a pie.

Example
The net assets of a firm consist of

	£
Buildings	25 000
Machinery	10 000
Motor vehicles	5 000
Investments	10 000
Inventories	5 000
Bank balances	5 000

It has no trade debtors or creditors. Its assets can be shown visually:

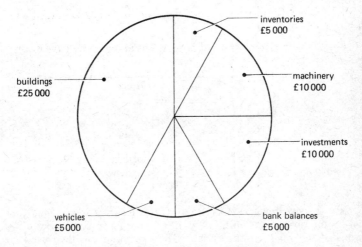

Circular test. A desirable property that is sometimes asked of index numbers (although rarely achieved) is that the index for year t based on year 0, say $I_t^{(0)}$, multiplied by the corresponding index for a later year s

based on year t ($I_s^{(t)}$), and divided by 100, should equal the index of year s based on year 0. That is, we require

$$\frac{I_t^{(0)} \times I_s^{(t)}}{100} = I_s^{(0)}$$

Neither the Paasche nor the Laspeyres index satisfies this condition, known as the circular test.

Circular triads.   If in comparing three objects A B C in pairs, A is preferred to B, B is preferred to C, and C is preferred to A, the triad A B C is said to be circular. If this occurs, it is not possible to rank the three objects since there is a basic inconsistency in the stated preferences.

City block metric.   A method of calculating distance between two points. The term is based on the idea of actual travel distance between two points rather than the shortest distance between them. The 'city block' is based on the fact that in American cities the streets tend to be at right angles to one another. The example below illustrates the way in which distance is measured using this metric. The shortest distance between A and B is represented by the dotted line AB. However, to get from A to B it is necessary to travel first to C (i.e. the distance AC) and then from C to B (i.e. the distance CB). Thus using the city block metric the distance from A to B is AC + CB.

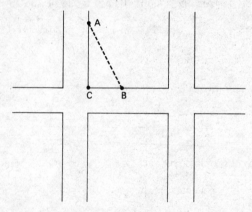

Class.   In specifying FREQUENCY DISTRIBUTIONS it is customary to group the observations into categories according to convenient divisions of the range of the variate. The categories are then used in subsequent analysis in order to make the work less laborious and/or to clarify the exposition. A category constructed in this way is called a class.

Class boundary. The values of the variable which determine the upper and lower limits of a CLASS are called class boundaries. For example, in a FREQUENCY DISTRIBUTION of age in a population the lowest age class might be 0–9 years. The class boundaries are then 0 years and 9 years.

Class frequency. The number of elements or individuals who fall in a particular CLASS is called the class frequency. In the table below, the class frequencies are 108, 56, 55, 131.

	Age	Frequency
Class 1	0–19	108
Class 2	20–29	56
Class 3	30–39	55
Class 4	40 and over	131
Total		350

Class interval. The range or interval between the highest and lowest values allowed in a particular CLASS. For example, if we classify football teams by the number of points they obtained in the Football League last year, we might have a frequency distribution like that given below:

No. of points	Frequency
Fewer than 25	5
25–34	24
35–44	25
45–54	20
55 and over	18
Total	92

Note that two of the class intervals are open – i.e. in the lowest class no lower limit is defined (although 0 could be assumed), and in the highest class no upper limit is defined. The other three classes are closed and the interval is 10 points.

Classification. For the purpose of convenient analysis, observations may be grouped into classes, so that the classes together exhaustively cover the entire range of values. The process of classification (i.e. the determining of CLASS INTERVALS, and consequently of CLASS FREQUENCIES) is an arbitrary process. Classes are humanly determined, whereas attribute categories (e.g. sex) are usually self-evident.

In contrast with this general meaning of classification, the term CLASSIFICATION STATISTIC means a statistic used to determine from which population a sample originated, when it may have been drawn from one of a number of POPULATIONS. In this context the term 'classification' assumes its more general non-statistical meaning of discrimination between one population and another.

Closed question.   A question in which the respondent is given a limited number of responses from which to choose. For example, the question 'Are you over thirty years of age?' presents only few choices – 'Yes', 'No', 'Don't know' and 'Refuse to answer'. In attitudinal questions, the choice is sometimes limited to a small number of categories. The respondent might be presented with a statement such as 'I approve of the way in which the Government is running the country' and asked to state which of the five categories – 'Agree strongly', 'Agree', 'Don't know', 'Disagree' and 'Disagree strongly' – was closest to his own position. (◆◆ OPEN-ENDED QUESTIONS and PRE-CODED QUESTIONS.) It is vital that the set of categories in closed questions should be both mutually exclusive (i.e. there should be no overlap between categories) and exhaustive (i.e. all possibilities should be covered by the alternatives). It may sometimes be necessary to include a residual category – 'Other, please specify' – in the CODING FRAME in case some unforeseen responses are received.

Cluster.   A group or collection of elements. The two principal senses of the word are: (1) in CLUSTER SAMPLING, where, instead of selecting a sample of elements directly, groups or clusters of elements are selected as, for example, by selecting a class of children in a school or a number of districts from a county where in each case the element or unit of analysis might be the individual person; and (2) in CLUSTER ANALYSIS, where the object of the exercise is to group similar elements together.

Cluster analysis.   A generic term used to describe a large body of statistical procedures which are designed to group, or 'cluster', either elements or variables in a data set according to some well-defined characteristic. The objective of the analysis may be: (1) data reduction, i.e. reducing the mass of data to manageable proportions. This is done: (a) by grouping similar elements together which will mean that it will be sufficient to consider the groups (of which there will only be a relatively small number) and not the elements (of which there may be a very large number) in subsequent analyses; or (b) by forming clusters of variables (in particular attitudinal questions) so that summary scores for the respondents on the clusters of variables may be used instead of having to use all the answers to the individual questions. (2) To construct a typology or classification of elements or variables – this may be appropriate in using sets of symptoms to construct a classification of illnesses for example. (3) To form classes or clusters of elements with particular characteristics – for example, identify deviant groups in a population. A form of cluster analysis has been used to identify the characteristics of drunken drivers. Such an analysis could be used for prediction purposes.

There are two basic kinds of cluster analysis: (i) divisive methods, which operate by splitting the original group up into successively smaller and more homogeneous groups; and (ii) agglomerative methods, which operate by linking together similar elements until the desired number of groups has been formed. The Automatic Interaction Detector (AID) is an example of a divisive method. Single linkage and complete linkage are examples of agglomerative methods.

The choice of SIMILARITY (or DISSIMILARITY) COEFFICIENT and of linkage method should be appropriate to the objectives of the analysis as widely different results can be obtained by different methods.

Cluster sample/sampling. In selecting a sample of students from a school system, it may be more convenient to select schools as the sampling units rather than individual students. Similarly in selecting a sample of industrial workers it may be desirable to use firms as the sampling unit rather than the individual worker. In both these cases, the sampling unit contains more than one population element; in other words, the sampling unit is a 'cluster' of elements. Sampling schemes of this kind are described as cluster sampling.

In order for cluster sampling to be satisfactory, each element in the population must be uniquely identified with one and only one of the clusters. If observations are taken of all the elements in the cluster, then the probability of selection of each element is equal to the probability of selection of the cluster to which it belongs. It is possible, however, to sub-sample elements for observation from each of the selected clusters – this is MULTISTAGE SAMPLING.

The principal advantage of cluster sampling over element sampling is that the cost per element is lower for cluster sampling due to lower cost of locating the elements. An example may illustrate this. In selecting a sample of schoolchildren in Greater London we could consider two sampling procedures. (1) An element sample for which we would require a complete list of all the schoolchildren in London: this would involve obtaining from every school a list of the children and then selecting a simple random sample from the combined lists; this, however, would be very expensive for two reasons – (a) the cost of the list and (b) the travel and other costs involved in interviewing or examining the selected children whom we could expect to be widely dispersed throughout a large number of schools. (2) Alternatively we could select a sample of schools from the list of schools and then interview all the children in the selected schools; this would reduce the cost by eliminating the need for the complete list of children (listing would only be necessary in the selected schools) and, since the selected elements would be closely grouped, the costs of the fieldwork would be greatly reduced. The same arguments apply to sample surveys of many other populations.

Cluster sampling does have disadvantages. By virtue of the fact that the selected elements are grouped, a sample of, say, twenty schools in which a total of 3000 children are examined may be much less representative of the population than 3000 children taken at random from the whole population of schoolchildren. This is because children within a school are likely to be similar to one another in many respects although they may be very different from children in other schools. By confining the sample to a few schools the estimate obtained may be much less precise than the estimate obtained from the corresponding element sample. In other words, the SAMPLING VARIANCE will be larger or the DESIGN EFFECT will be greater than 1. How seriously the precision is affected depends on the INTRA-CLASS CORRELATION COEFFICIENT in the population – the degree to which elements within clusters are similar to one another and different from elements in the other clusters.

Cluster sampling should be preferred when the advantage of lower cost per element outweighs the disadvantages of lower precision per element and higher cost of analysis.

Cluster size. The number of elements in a cluster. In human populations, most clusters are of unequal size, e.g. schools, firms, towns. However, in some cases clusters of equal size can be found, e.g. the number of index cards in a file may be constant, or the number of dwelling units per block in a new housing development.

Cochran's criterion. A criterion proposed by W. G. Cochran to test for significant differences between three or more sets of matched observations when the observations are divided into two categories, such as good or bad, unfavourable or favourable.

The relevant statistic Q is calculated using the formulae

$$Q = \frac{(c-1)\,(c\Sigma X_j^2 - (\Sigma X_j)^2)}{c(\Sigma Y_i) - (\Sigma Y_i^2)} \text{ or } \frac{c(c-1)\,\Sigma\,(X_j - \overline{X})^2}{c(\Sigma Y_i) - (\Sigma Y_i^2)}$$

where c is the number of sets of matched observations (arranged as columns) and X_j the total successes in the jth sample and Y_i the total successes in the matched group or row i.

Example

A statistical formula is taught to students and tested by using three questions, 1, 2 and 3. The students who succeed in each question are listed in the following table. The total correct answers for each student are shown in column Y_i, and the squares of the totals in column Y_i^2. Is there significant difference in success on the questions?

student \ question	1	2	3	Y_i	Y_i^2
1	X	X		2	4
2	X	X	X	3	9
3		X	X	2	4
4		X	X	2	4
5	X	X	X	3	9
6	X	X	X	3	9
7	X		X	2	4
8		X		1	1
	$X_1 = 5$	$X_{i_2} = 7$	$X_3 = 6$	$\Sigma Y_i = 18$	$\Sigma Y_i^2 = 44$

There are three groups, $(c - 1) = 2$, and totals from 1, 2 and 3 are squared

$$Q = \frac{2 \times \{[3 \times (25 + 49 + 36)] - 18^2\}}{(3 \times 18) - 44}$$

$$= \frac{12}{10}$$

$$= 1.2$$

If, using two degrees of freedom (i.e. $(c - 1)$), a chi-squared table is consulted, the statistic 1.2 is not significant even where $P = 10$ per cent. We conclude that there is no significant difference between the students' successes on the three questions.

Cochran's Q. The statistic used for COCHRAN'S CRITERION representing an adaptation of the chi-squared statistic test for significant differences between three or more sets or samples of matched observations.

Code. A form and language into which a set of instructions have to be translated, or in which a set of instructions must be written for computer operation. CODING SHEETS, usually consisting of 80 character spaces per line, are provided for programming.

Alternatively, the word is sometimes understood to mean ('1) a numerical notation, such as the BINARY notation or (2) a rule used for translation of random numbers into a form usable for voucher numbers where, for example, the system of voucher numbers adopted, uses a mixture of alphabetical and numerical characters.

The term code also denotes a number (a code number, or code) which is used to identify a category or class for categorized data. For example, in the frequency distribution below the numbers in the final column are the codes used to identify the classes in analysing the data.

Class	Frequency	Code
Under 15	78	1
15 and under 16	91	2
16 and under 17	65	3
17 and under 18	83	4
18 and over	103	5

Code book.   The full set of CODING FRAMES which cover all the responses to an interview schedule or questionnaire is called the code book for the questionnaire. The code book is in effect the dictionary of the data set and without it the summarized data on the coding sheets, punched cards or magnetic tape are meaningless.

Coding.   Once the initial editing has been completed, the next operation is to prepare the data for analysis. In most surveys this involves coding the responses. The procedure of coding involves three stages. The first stage is to decide how to categorize the responses; the second stage is to allocate numerical or symbolic values to the categories; the third stage is to allocate each individual response to the appropriate category. The principal purpose of coding is to provide a convenient way of summarizing the responses (or observations) in surveys so that the data can be analysed more easily, generally on a computer. The set of categories for a question or set of questions, together with the numerical or symbolic codes to be allocated to them, is called a CODING FRAME.

Coding may in effect be done before the fieldwork is carried out by printing the categories of response and the codes on the schedule itself. In this case all the interviewer (or in a self-completion questionnaire, the respondent) needs to do is to ring the appropriate code. Otherwise the coding must be done at a later stage.

Coding error.   Any errors which arise in the CODING of the responses. The degree to which errors arise will depend on the ability of the coders, on the suitability and completeness of the CODING FRAME and on the degree to which the editing has ensured the completeness and accuracy of the responses.

Coding errors form a component of NON-SAMPLING ERRORS, but if the coding operation is kept simple and careful checking is carried out at all stages the problems should not be too severe. The evidence on coding errors suggests strongly that complex combination codes should not be used during the coding operation but should be constructed at a later stage during the analysis using the computer, thus making the task facing the coders as simple as possible.

Coding frame. The set of categories into which we classify the responses to a question or other stimulus, together with the numerical or symbolic codes we allocate to these categories, is called a coding frame. In many cases, the categories are self-evident and no serious problems arise. For example, to the question 'Are you over forty-five years of age?' there are only four answers: 'Yes', 'No', 'Don't know' and 'Refuse to answer'. Thus the coding frame is easy to construct. Any set of numbers can be allocated to the possible answers, e.g. 'Yes' = 1; 'No' = 2; 'Don't know' = 7; 'Refuse to answer' = 8. In many cases the situation is not so straightforward. If the question were 'What is your occupation?' the number of possible replies is extremely large. Here the researcher is faced with a dilemma – using too many categories will be wasteful both because responses may be difficult to allocate to specific categories and because the analysis may well become unyielding and expensive; on the other hand, using too few categories will discard a lot of useful information and may lead to misleading results. A possible solution here would be to use the five classes: Professional = 1; Managerial = 2; Other non-manual = 3; Skilled manual = 4; Other manual = 5; with additional codes – Not asked = 6; Don't know = 7; Refuse to answer = 8; and Not Applicable = 9. It should be borne in mind, however, that the number and type of categories will depend on the purposes for which the analysis is to be carried out.

OPEN-ENDED QUESTIONS cause special problems since it is very difficult to predict in advance what the responses will be. The researcher may well set up in advance a list of categories which he thinks will be appropriate but that list will almost certainly have to be modified once the responses have been examined. It is often a good idea to examine a sample of the responses before the main coding operation commences in order to construct a suitable coding frame. Even when this is done, the frame may need to be modified further if important new categories of response are found during the actual coding.

One minor point to note: for convenience in the analysis it is desirable to use the same codes consistently for the ever-present categories Don't know, Refuse to answer and Not applicable. As in the examples above, the codes 7, 8 and 9 – or 77, 88, 99 – may be used for this purpose.

Coding sheet. The form on to which the codes for the responses to a survey schedule are written in preparation for punching. In some cases, where all the questions have pre-coded responses, the punching may be done directly from the questionnaire or schedule.

Coefficient. Literally, a value that 'acts [or is employed] together with' another value. In mathematics, the term usually means a scalar constituent of a term used in a multinominal expression. Thus, for example, in the expression $5x + 9y + 15z$ the values 5, 9 and 15 are said to be the coefficients of x, y and z respectively. In statistics the word has a wider meaning than its general mathematical meaning. Thus, although the term 'regression coefficient' denotes the value b in, for example, the equation

$$y = a + bx$$

where it performs a function similar to coefficients in mathematics generally, the term 'correlation coefficient' on the other hand, means a scalar value to be understood, not in the context of variables, such as x, y and z, but rather in the context of a scale of measurement that varies between -1 (perfect negative correlation) and $+1$ (perfect positive correlation).

Other coefficients, such as those of association (Yule's coefficient), and those of concentration (the Gini coefficient) should also be understood in the context of an accepted, comparative, scale used for the measurement of a particular quality, such as correlation, association or concentration.

β-Coefficient. ◗ BETA-COEFFICIENT.

Collapsed strata/collapsed stratum method. It may happen that in order to utilize all the prior information we have about the population that we STRATIFY to the point where we select only one unit from each stratum. It is therefore impossible to estimate the SAMPLING VARIANCE directly since we would need at least two units from each stratum to do this. We can use the method of collapsed strata to cope with this problem. We first form pairs of strata on *a priori* grounds – forming collapsed strata – and then calculate the variance treating these collapsed strata as single strata each of which contains two sample units. By ignoring the finer stratification we will tend to overestimate the variance but this overestimation should not be too important, especially when we are dealing with SUB-CLASSES.

Colligation/coefficient of colligation, Yule's coefficient of colligation. Literally 'bound together'. This term is similar in meaning to ASSOCIATION. The coefficient of colligation measures the relationship between two attributes in much the same way as the various COEFFICIENTS OF ASSOCIATION.

The coefficient of colligation defined by Yule is:

$$Y = \frac{1 - \sqrt{\left(\frac{bc}{ad}\right)}}{1 + \sqrt{\left(\frac{bc}{ad}\right)}}$$ using the category symbols

a	b
c	d

where a, b, c and d are the four categories in a two-way table.

Examples

1. Of the thirty-two students who attempt two examinations the results are as follows:

	Pass Exam A	Fail Exam A
Pass Exam B	16	0
Fail Exam B	0	16

The coefficient of colligation $Y = \dfrac{1 - \sqrt{\left(\frac{0}{256}\right)}}{1 + \sqrt{\left(\frac{0}{256}\right)}}$

$$= 1$$

2. Of the thirty-two students who attempt two examinations the results are as follows:

	Pass Exam C	Fail Exam C
Pass Exam D	4	12
Fail Exam D	12	4

The coefficient of colligation $Y = \dfrac{1 - \sqrt{\left(\frac{144}{16}\right)}}{1 + \sqrt{\left(\frac{144}{16}\right)}}$

$$= \frac{1 - 3}{1 + 3}$$

$$= -0.5$$

In the first case the coefficient is the same as that for COEFFICIENT OF ASSOCIATION but generally the results are different, as we can see from the second example.

Column percentage. The percentage of any particular column in a given cell. For example, if the column total is 500 and the particular cell value is 25, its column percentage is 5.

Column total. The total of all elements in a column (i.e.

$$\sum_{j=1}^{i} a_j$$

where $a_1, a_2, \ldots a_i$ are each of the elements of the column).

Combination. A set of characters, objects or events which ignores the order in which they occur, as distinct from PERMUTATION, which takes the order of their occurrence into account.

For example, let us suppose that a coin is spun three times. There are eight possible permutations indicated below.

Row	First throw	Second throw	Third throw	Combination
1	Head	Head	Head	HHH
2	Head	Head	Tail	HHT
3	Head	Tail	Head	HHT
4	Head	Tail	Tail	HTT
5	Tail	Head	Head	HHT
6	Tail	Head	Tail	HTT
7	Tail	Tail	Head	HTT
8	Tail	Tail	Tail	TTT

In this case the two sub-sets or classes are heads and tails, and each class has a number of outcomes (e.g. 3 heads: 0 tails; 2 heads: 1 tail). The combination column places heads first, and it can be seen that the combination HHT can occur in three ways, or permutations. The permutations are the orders head, head, tail, as in row 2; head, tail, head, as in row 3; or tail, head, head, as in row 5.

Combinatorial. A term which denotes the number of different COMBINATIONS of size r units which can be drawn from a set of n units. This quantity is used in calculating the coefficients of the binomial expansion and in computing the probability of selection of a particular sample in simple random sampling without replacement. It is written either

^nC_r or $\binom{n}{r}$ where r is the smaller number and n is the larger group. The formula for the calculation of combinatorials is

$$\frac{n!}{(n-r)!\,r!}$$

which may be simplified to

$$\frac{n \times (n-1) \times (n-2) \times \ldots (n-r+1)}{1 \times 2 \times 3 \times \ldots r}$$

For example, the number of pairs which can be drawn from five elements A, B, C, D and E can easily be deduced

AB,AC,AD,BC,BD,AE,CD,BE,CE,DE

This number of pairs can be computed by applying the formula where n = 5 and r = 2. Thus ^nC_r or $\binom{n}{r}$ is

$$\frac{1 \times 2 \times 3 \times 4 \times 5}{(1 \times 2 \times 3) \times (1 \times 2)} = \frac{120}{6 \times 2}$$

$$= 10$$

But as $(1 \times 2 \times 3)$ is common to both numerator and denominator the formula can be simplified to

$$\frac{5 \times 4}{1 \times 2} = 10$$

Thus, in this case, the number 10, which was obtained by writing out the combinations, can be calculated (or verified) by using the formula.

Combined strata/combined stratum method. When dealing with paired selection SAMPLE DESIGNS, the formula for calculating the sampling variance is very simple. However, NON-RESPONSE or the fact that we are dealing with SUBCLASSES may destroy this simplicity. One way of dealing with this problem is to use the method of combined strata. We take one random selection from each pair and treat the combined set as one combined selection. The remaining selection from each of the strata forms the second combined selection. These two combined selections are then used as the basis for the calculation of the sampling variance. The calculation is considerably more complicated but at least it preserves the gains of the stratification built into the design. The combined selections should be based on at least five to ten strata.

Communicate. A description of the relationship between two states in a Markov chain where they are mutually accessible from each other, that is, if A affects B and B conversely affects A.

For example, in the entry BRANCHING PROCESS the state 'customers' affects 'accounts paid' (0.5) and 'accounts paid' also affects 'customers'

(0.8). Irrespective of other consequent states of both, these two states affect each other, or communicate.

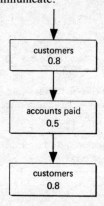

Complete coverage.   A survey or census can be described as complete if all of the elements in the population are included in the enumeration. Such a survey or census involves complete coverage of the population.

Complete linkage/farthest neighbour.   A form of CLUSTER ANALYSIS. As in SINGLE LINKAGE analysis the two most similar elements are first linked together. However, in order that another element be added to a cluster, the additional element must be sufficiently similar to all the elements already in the cluster. Thus the method will produce compact clusters, such that each of the elements in a cluster is closely related to (similar to) all the other elements in the cluster. The method is sometimes referred to as 'farthest neighbour' analysis since whether an element is included in a cluster depends on its similarity to the element already in the cluster which is most different (distant) from it.

Consider the example of a similarity matrix given in the entry SINGLE LINKAGE. There are seven elements and we wish to form two clusters. Unless the elements in the data set are closely related in homogenous groups, it may be difficult to construct satisfactory clusters using this method. If the data set looks like diagram 1, all will be well and the two indicated enclosed clusters will be formed.

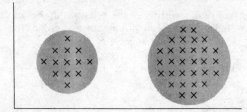

Diagram 1

However, if the data set looks like diagram 2, SINGLE LINKAGE might be much more appropriate. The choice of method depends, of course, on the objectives of the analysis, and great care should be taken to ensure that the choice both of similarity measure and of linkage method is appropriate to the problem.

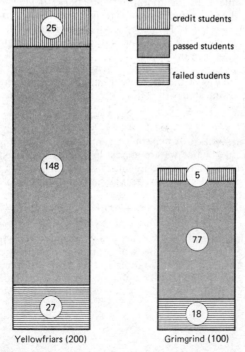

Diagram 2

Component bar chart. A special kind of BAR CHART used for comparing a number of attributes in each of a number of discontinuous or classified categories or interval groups.

If we use the Yellowfriars/Grimgrind example cited in the CHI-SQUARED TEST entry, the bar chart drawn from the figures of examination results would take the following form:

credit students

passed students

failed students

Yellowfriars (200)   Grimgrind (100)

If there had been more than two schools, the chart would consist of a larger number of bars. Similarly, each bar can consist of more than three components.

Components of variance.   The determination of components of variance is one of the functions and objects of ANALYSIS OF VARIANCE. For example, assume that a workshop contains five operatives, all engaged in the manufacture of an identical product. Figures are available which show the number produced by each operative on each of five days of a given week. There are thus twenty-five values, of which five each constitute a sample for each operative and five each constitute a sample for each of the days of the week.

There are 24 (i.e. 25 − 1) degrees of freedom for the total sum of squares of 25, and the mean square (variance) can be estimated by dividing the total sum of squares by 24. The sum of squares consists of a number of components, for example, the INTERCLASS or BETWEEN-GROUPS SUM OF SQUARES such as between operatives or between days and the INTRACLASS or WITHIN GROUP SUM OF SQUARES, in this case within the cells of the two-way table. From each of the sums of squares, the corresponding mean square (variance) can be obtained by dividing the sum of squares by the appropriate number of degrees of freedom. The ratio of the mean squares can be tested using the F-TEST or F-RATIO TEST at a chosen level of significance. A significant value of the test statistic would indicate a significant difference, for example, between the output of the five operatives.

The concept is extended in MULTIPLE REGRESSION ANALYSIS, where the sum of squares of the dependent variable is separated into those components associated with regressors and that associated with the residual. (◗ VARIANCE, ANALYSIS OF.)

Composite hypothesis.   A hypothesis composed of a number of simple hypotheses. In practice we are almost invariably interested in tests of composite rather than simple hypotheses. We may wish, for instance, to test the simple NULL HYPOTHESIS that a particular model of car has a mileage rate per gallon of 35 against the composite alternative hypothesis that the mileage per gallon is less than 35.

Composite index number.   Correctly any index number which is composed by aggregating the values of a number of heterogeneous elements. For example, if bread, butter and tea are considered good representative commodities for compiling a cost of living index, then an index compiled using the weights below would be a composite index number.

	Price	*Weight*	*Product*
Bread	20	5	100
Butter	25	2	50
Tea	30	1	30
			180

What constitutes a set of heterogeneous elements or items will, however, depend on the objective of compilation of the index. For example, the *Financial Times* Ordinary Share Index would be regarded as composite from the standpoint of one company since it is composed of share prices of a number of leading companies. It would not, however, be regarded as composite when considered as an index of equities as against prices of other forms of investment.

Concentration. The extent of disproportion in the dispersion of some variables (e.g. the concentration of low income as against high income frequencies). The extent of concentration may be measured by means of the LORENZ CURVE or the GINI COEFFICIENT. (◆◆ CONCENTRATION, INDEX OF.)

Concentration, index of. An index designed to measure CONCENTRATION at any point on a scale. If $x_1, x_2, \ldots x_n$ are the values of a variate (such as income or wealth) and $f_1, f_2, \ldots f_n$ are the frequencies of those values (such as the number of persons earning income x_i), then the index of concentration at any given point $(k + 1)$ on the scale is the value C, where

$$\left[\frac{\sum\limits_{i=n-k+1}^{n} f_i x_i}{\sum\limits_{i=1}^{n} f_i x_i} \right]^{C} = \frac{\sum\limits_{i=n-k+1} f_i}{\sum\limits_{i=1}^{n} f_i}$$

The left term is the sum of products of the last $n - k$ frequencies and values expressed as a fraction of all products of frequencies and values. The right term is the cumulative sum of frequencies expressed as a ratio in terms of the sum of all frequencies. The index of concentration is the power to which the left term is raised to equal the right term.

This measure should not be confused with the GINI COEFFICIENT, which is a general measure of concentration. The above index indicates concentration at a particular point on the curve. (◆ CONCENTRATION.)

Concordance.   Unlike CORRELATION, the term concordance is usually confined to positive agreement between ranked variates. It corresponds to positive correlation in the case of quantitative variates, and is its equivalent in rank correlation.

Concordance, coefficient of.   A coefficient which particularly measures the extent of concordance, and is concerned with the extent of agreement among a number of ranked sets irrespective of paired comparison, unlike the COEFFICIENT OF AGREEMENT, which is based on numbers of pairs of identical rankings.

   The formula is

$$C = \frac{12\Sigma d^2}{m^2 n(n^2 - 1)}$$

where d is the deviation of the sums of rankings from their common mean $[\frac{m (n + 1)}{2}]$, and where m is the number of rankings and n the number of objects.

Examples
If we take the orderings ABC and ACB we have $n = 3$ and $m = 2$. The sums of ranks are calculated, using $A = 1$, $B = 2$ and $C = 3$.

Rank (1)	1	2	3
Rank (2)	1	3	2
Sums of ranks	2	5	5
Common mean	4	4	4
$d =$	-2	1	1

$\Sigma d^2 = 4 + 1 + 1$
     $= 6$

The coefficient of concordance $C = \dfrac{72}{96}$

                                   $= 0.75$

With the orderings, (ABC, ACB, CAB, BAC), the common mean of rankings is 8, and $\Sigma d^2 = 6$, but $m = 4$ and $n = 3$. Thus

$C = \dfrac{72}{384}$

   $= 0.1875$

In the case of complete concordance, i.e. (ABC, ABC), because the common mean of $(1 + 1, 2 + 2,$ and $3 + 3)$ is 4, therefore $\Sigma d^2 = 8$. Thus

The coefficient of concordance $C = \dfrac{96}{96}$

$$= 1$$

Concurrent validity. If a scale is constructed as an indicator of behaviour or of some independently observable criterion, its validity may be measured by the degree to which it succeeds in measuring this criterion. When the alternative measurement is carried out at about the same time, this is called concurrent validity. If the other measurement or observation takes place in the future, it is PREDICTIVE VALIDITY.

Conditional probability. The probability of some event occurring given that some other event on which it is dependent, or with which it is connected, has already occurred. For example, in a group of 100 people, 50 are regular readers of the *Daily News*, and 30 are supporters of unilateral nuclear disarmament for Britain. There is some overlap in that 10 are both, but also that 30 are neither. The probability that a person selected at random from the group will be a supporter of unilateral disarmament is 3/10. However, when other sources are checked it may be known that the selected person is a reader of the *Daily News*. Although the same experiment has been performed, additional information is now available. We desire the probability that the selected individual is a supporter of unilateral disarmament knowing that he is a *Daily News* reader. Clearly the person must be one of the 50 *Daily News* readers and, out of these, 10 are also supporters of unilateral disarmament. The conditional probability is then 1/5. Having the information that the person is a *Daily News* reader has reduced the probability of his being a supporter of unilateral disarmament.

There are some situations in which these two probabilities will be the same, and here the two events under consideration are said to be independent. For example, if a card is drawn at random from a well-shuffled deck, then replaced before a second card is drawn, then the outcome of the latter experiment is independent of the outcome of the first.

A particular application of the concept of conditional probability can be found in BAYES' THEOREM.

Conditional probability distribution. If the variates $x_1, x_2, \ldots x_p$ have a joint probability distribution, the distribution of a subset of variates holding the others constant is called a conditional probability distribution. (◗ CONDITIONAL PROBABILITY.)

Confidence interval.   The interval (or difference) between two limits L_1 and L_2, which are called the confidence limits. It is estimated at a particular CONFIDENCE LEVEL (for example 95% or 99%) that the interval contains the value of a parameter.

For example, if a simple random sample of size n is drawn from a large population, and the sample mean is $\overline{X}$ and the standard deviation is s, the 95% confidence interval estimate of the population mean is $\overline{X} \pm 1.96\, s/\sqrt{n}$. The CENTRAL LIMIT THEOREM tells us that of all interval estimates of this form for a particular sample size n, 95% will include the value of the population mean.

Confidence level.   The probability (expressed usually as a percentage) that a CONFIDENCE INTERVAL will include a particular parameter value. For example, 95 and 99 per cent are common confidence levels. (♦♦ CONFIDENCE LIMITS.)

Confidence limits.   The limits L_1 and L_2 which constitute the upper and lower boundaries of a CONFIDENCE INTERVAL. (♦♦ CONFIDENCE LEVEL.)

Confounding.   A term used in the design of experiments to describe a situation where the study of less important treatments and interactions is sacrificed so that more important treatments and interactions can be studied fully. This may involve combining some of the less important treatments and interactions and dealing with them as a single group. An example is the design of experiments involving a number of blocks where the size of blocks may be limited so that not all possible treatment combinations can be studied in each block. Thus the number of interactions which can be studied must be reduced.

Although confounding is usually part of the design, it may occur accidentally or naturally (e.g. where it is not possible to study each of a number of interactions individually).

Consistence, coefficient of.   When the ranking of a number of objects is being investigated by the method of PAIRED COMPARISONS, the consistency of the judgements expressed can be measured in terms of CIRCULAR TRIADS. If the number of objects involved is n and the number of circular triads observed is d, the coefficient of consistence is

$$c = 1 - \frac{24d}{n^3 - n}$$

if n is odd, and

$$c = 1 - \frac{24d}{n^3 - 4n}$$

if n is even.

Consistency.   A term which can be used in two contexts.

First, consistency is concerned with the internal agreement of data or procedures among themselves. In EDITING data, for example, we can compare different sources of information to see whether they are in agreement (i.e. are CONSISTENT with one another). The same sense is involved in the COEFFICIENT OF CONSISTENCE, which is a measure of the degree to which a set of paired comparisons is consistent. For example, if an individual prefers a to b, prefers b to c, but prefers c to a, he is being inconsistent.

The second use of the term is in CONSISTENT TESTS or CONSISTENT ESTIMATORS.

Consistency checks.   A part of the editing process in preparing survey data for analysis is to check whether the answers given to different questions are consistent with one another. One of the most common forms the checks take is to ask for the same information in two or more different ways. If, for example, at the beginning of the interview the respondent is asked for his date of birth, he might be asked at the end to give his age. The two answers should agree. Similarly if age, age at marriage and number of years of marriage are all obtained during the interview, then the number of years of marriage should be equal to age less age at marriage. The difficulty is that if the checking is done after the interview is completed and a discrepancy appears, there is no way of telling which of the answers is correct. However, the checks are valuable in providing an indication of the quality of the information collected.

Consistent.   Compatible with, not contradictory. The term is used in two contexts: the consistency of responses given by a respondent (▶ CONSISTENCY CHECKS), and the consistency of estimators and tests.

Consistent estimators.   An estimator of some population parameter based on a sample of size n will be consistent if its value gets closer and closer to the true value of the parameter as n increases. In a loose sense, whatever particular sample is selected we know that the estimator will give us an estimate of the parameter as close to its correct value as we please, provided that the sample size is sufficiently large. An estimator will be inaccurate to the extent that it gives values of the estimate different from the true parameter value. The desirable property of consistency ensures that this 'inaccuracy' decreases as n increases.

For example, an estimator $\hat{\Theta}_n$ computed from a sample of n observations is consistent if, for any positive ε and η, however small, there is some N such that the probability that $|\hat{\Theta}_n - \Theta| < \varepsilon$ is greater than $1 - \eta$ for all $n > N$. If $\hat{\Theta}_n$ is a consistent estimator of Θ it is said to converge in probability to Θ in this sense.

Consistent tests. Test procedures which reject false null hypotheses with increasing probability as the sample size n on which the test is based increases. It is reasonable to ask of a test that its accuracy in drawing correct inferences increases as the amount of information, in the form of a larger number of observations, increases. The probability of committing a TYPE II ERROR decreases as n increases, such that we can make the type II error for any true member of the compound alternative hypothesis as small as we please, provided we choose a sufficiently large n.

Constant. A quantity which remains unchanged. The opposite of a variable.

Constraint. In the case of data, a constraint is a limitation imposed by external conditions. In LINEAR PROGRAMMING, it is a limitation imposed by a fixed quantity of a resource.

For example, suppose that two units of a raw material are used in the manufacture of product x_1 and five units of the same material are used in manufacture of product x_2. The total number of (say, 100) units available imposes a constraint on usage. In linear programming this would be written

$$2x_1 + 5x_2 \leq 100$$

Construct validity. On the basis of theoretical considerations, the investigator may be able to postulate the direction and magnitude of the relationships between a scale and other variables being measured. The degree to which the observed relationships conform to the postulated relationships is a measure of the validity of the scale – this is construct validity. Failure of the observations to conform to the expectation may well indicate that the theoretical basis is inappropriate and not that the scale is invalid. Agreement between the two will, however, give the investigator stronger confidence in the scale, and disagreement should lead to a careful consideration of alternatives.

Consumer's risk. The risk of accepting the NULL HYPOTHESIS when the alternative is true, similar to BETA-ERROR. In the specific context of consumer and producer it is the risk that an unsatisfactory batch will be accepted by a given sampling plan, a risk which depends on the level of quality required and on the criteria of the sampling plan.

Contamination. Doubtful accuracy or inaccuracy which arises because one observation has been affected by another or others. The term is sometimes used in data preparation (e.g. where one record is affected by another) but the main usage of the term is in interviewing, where, for example, the report on an interview with B will be coloured (i.e. contaminated) because it has been affected by the interviewer's previous interview with A in the same household.

Content validity. A more systematic approach to the validity of a scale than FACE VALIDITY. In this case the items in the scale are examined to ensure that: (1) they are all concerned with the attitudinal continuum under study; (2) that the whole range of the attitude is covered by the items; (3) that no particular aspect of the attitude is given undue weight. If all three conditions are satisfied, the scale has content validity for that attitude. The whole process involves subjective judgement and may best be delegated to a panel of experts rather than carried out by the investigator himself.

Contingency coefficient. A coefficient which attempts to measure the degree of association or dependence between two variates as expressed in a CONTINGENCY TABLE. In a general sense, contingency and ASSOCI-ATION are identical, although association is sometimes more narrowly defined as the dependence in a 2 × 2 table.

Pearson's coefficient of contingency is given by

$$c = \left(\frac{\chi^2}{n + \chi^2} \right)^{1/2}$$

where χ^2 is the CHI-SQUARED STATISTIC and n is the number of observations.

Contingency table. The elements in a population or group may be classified according to qualitative (categorical) variables (for instance, sex, race, religion or political party). A classification in a two-way table of the elements according to two such qualitative characteristics is called a contingency table. The rows of the table denote the categories of the first variable, the columns the categories of the second variable. Each cell in the table contains the number of individuals having both the characteristic of the row and column of the table to which the cell belongs. Thus in the table below the cell in the second row of the third column contains 100 individuals. These 100 individuals have the characteristic of the second row (i.e. they are Catholics) and the characteristic of the third column (i.e. they voted for a party other than the Democrats or the Republicans).

religion＼party	Democrat	Republican	other	total
Protestant	180	300	120	600
Catholic	240	170	100	510
other	95	30	60	185
total	515	500	280	1295

Continuity, continuous.   A variate is continuous if it may take values in a continuous range.

A probability or frequency distribution is continuous if it relates to a continuous variate, i.e. the distribution of a continuous variate. In standard notation is denoted by

$$f(x)\mathrm{d}x : -\infty < x < +\infty$$

where the range of the continuous distribution is infinite; and by

$$f(x)\mathrm{d}x : a \leqslant x \leqslant b$$

where the range is finite.

It is possible to represent a continuous distribution by a discontinuous one, by dividing its frequency into categories based on interval measurements. Thus, for example, an age distribution, which is theoretically continuous, is frequently represented in terms of the frequencies in five-year age groups.

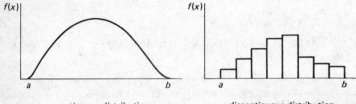

continuous distribution   discontinuous distribution

Continuous sampling (sequential sampling).   An extension of acceptance sampling and double sampling techniques. Using a given maximum number of defectives or maximum rate of occurrence of defects (x) and a predetermined confidence level, it is possible to calculate an acceptance number and a rejection number for any size of sample. As the size of sample increases, so will the acceptance and rejection numbers. If at any time the number of defects is less than the acceptance number, the sample is accepted. If the number of defects is greater than the rejection number, the sample is rejected. Otherwise sampling is continued. (◆◆ ACCEPTANCE SAMPLING, CONTROL CHART, CONTROL LIMITS.)

Control chart.   A graphical method of controlling size of product or quantity of input or output in industrial production. A recommended standard is prescribed, for example, that a screw should be 2 cm long or that a tablet should contain 300 mg of Aspirin. Tolerable variances are calculated in each case, and it may, for example, be decided that the recommended standard deviation, denoted by σ, in length in the case of 2 cm screws should be 0.0025 cm. For the purpose of control, warning limits and action limits are prescribed within which differences can be tolerated. If the limits of 1.96σ and 2.58σ are used, warning limits will be drawn at 2 cm $\pm$ 1.96 $\times$ 0.0025 cm, and action limits at 2 cm $\pm$ 2.58 $\times$ 0.0025 cm.

If a single screw taken from a sampling procedure measures more than the upper action limit, or is smaller than the size prescribed by the lower action limit, there is cause for immediate investigation. Alternatively, if a number of screws successively have measurements in the intervals between action and warning limits, there is need to investigate for bias in the manufacturing process, since the equipment may be producing consistently oversized or undersized batches of screws.

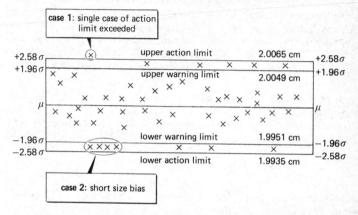

Note that in practice it is common to examine batches rather than single items (such as screws). In this case, the control limits for the batch mean are

$$\pm \ \frac{1.96\sigma}{\sqrt{n}} \ \text{ or } \pm \ \frac{2.58\sigma}{\sqrt{n}}$$

where n is the size of the batch. ($\blacktriangleright\blacktriangleright$ CONTINUOUS SAMPLING, STANDARD DEVIATION, STANDARD ERROR.)

Control group.   In experimental testing a group of persons or objects used as a standard of comparison or accepted norm with which to evaluate others among which a new process or method, or set of processes and methods, has been implemented. For example, in testing the effects of two drugs A and B for quick recovery among a set of patients, three groups of patients would be used to carry out the experiment: (1) those injected with doses of drug A; (2) those injected with drug B; and (3) those injected with a neutral substance and kept under similar group conditions as those of groups (1) and (2). Whereas the results of experiments with groups (1) and (2) can be compared with each other, group (3) provides a standard of measurement for both.

Control limits.   Limits used in quality control for prescribing tolerances or allowable deviations from a required standard size in measurement or quantity in input and output. A given process may indicate that it is possible to manufacture a match with a length of 4 cm and STANDARD DEVIATION σ of 1 mm. Packing problems require that there should be an upper limit, which should not be exceeded; and user considerations (e.g. holding time after striking a match) predetermine a lower limit of size. For the purpose of control, warning and action limits are often prescribed at 1.96σ and 2.58σ respectively. In this example, if a single match exceeds 4.258 cm or is less than 3.742 cm the manufacturing process will undergo immediate investigation. If, alternatively, there is frequent successive transgression of the warning limits (4.196 cm and/or 3.804 cm) steps will be taken to discover why short (or long) matches are consistently being produced. (◗ CONTROL CHART, QUALITY CONTROL.)

Controlled process.   A production process subjected to valid quality control procedures so that the size, input content or output requirements of a given product are statistically stable. The means and variances of any or all of these variables are consequently predetermined. Any variation in size, input content or output requirements of the product will be the result of random effects.

Correlation.   In the widest sense this term means any kind of association or interdependence between sets of data, whether of (1) ATTRIBUTES, (2) ORDINAL measurements or (3) quantitative measurements, such as height, weight, length, pressure or cost. In a stricter sense, the term applies to association or interdependence between measurements of variables, as in (2) and (3) above. It is measured by means of COEFFICIENTS, such as the Pearson PRODUCT MOMENT CORRELATION coefficient, and KENDALL'S TAU or SPEARMAN'S RHO, the latter two being used for ordinal measurement. The measurements of correlation usually scale between -1 (i.e. perfect inverse correlation) and $+1$ (perfect positive correlation). If there is no association whatever, the coefficient is 0.

It is not necessarily true that positive correlation between two variates proves them to be dependent on each other. The correlation may have resulted from mutual dependence on a third (unstated) variate. For example, if ice-cream sales rise, sales of beachwear also rise. The increase in sales of beachwear does not cause an increase in sales of ice-cream, nor does the increase in sales of ice-cream cause an increase in sales of beachwear. They would probably be competitive expenditures, for both are means of keeping cool. The positive correlation between the two sets of sales figures results from their being dependent on weather conditions. If weather becomes warm and congenial, both beachwear sales and ice-cream sales are likely to rise.

Correlation coefficient. A measure of CORRELATION between two variates. The most common measure of correlation between quantitative measurements is the Pearson PRODUCT-MOMENT CORRELATION (ρ or r), which takes values ranging from -1 (perfect negative correlation) and $+1$ (perfect positive correlation). The measure (ρ) of correlation between two sets of values $(x_1, x_2, \ldots x_i)$ and $(y_1, y_2, \ldots y_i)$ is indicated by the formula:

$$\rho = \frac{\sigma_{xy}}{\sigma_x \cdot \sigma_y}$$

where σ_{xy} is the covariance of x and y and σ_x and σ_y are the standard deviations of x and y respectively.

The PRODUCT-MOMENT CORRELATION is so called because its numerator is the COVARIANCE (also termed the first product-moment) of the two variates concerned.

SPEARMAN'S RHO is the product-moment correlation coefficient between ordinary rankings. KENDALL'S TAU is a different correlation coefficient for use with ranked data.

Correlation, spurious. The freak correlation of random observations of generally uncorrelated variates. The term is used to denote a condition which rarely occurs in practice, where, using a random method of selection from a series of pairs of values from two completely unassociated and uncorrelated variates, the series of observed values appear to be highly correlated. This condition should not be confused with nonsense correlation, (or illusory correlation), where the correlation between two variables can be explained by reference to a third variable but not by reference to each other, e.g. sales of ice-cream and sales of beachwear.

Cost function. The statistical meaning of this term (as distinct from its meaning in economics and accounting) is the function which gives the cost of obtaining a particular sample as a variable, dependent on the relevant factors which affect cost. A point of local optimality may be reached where the marginal cost of obtaining additional sample material is equal to the added marginal value of any additional precision arising from larger sample size.

Cost of living index. This is a somewhat misleading term that is applied to indices of retail or consumer prices. Sometimes it refers to the 'average' family and the rate by which its income has to increase in order to maintain its standard of living. In other uses a more limited scope is intended. For example, a cost of living index was instituted in Great Britain in July 1914 to measure the cost of maintaining the standard of living of working-class households, and as such it measured essentially the minimum cost of living. This has since been replaced by a retail prices index whose quantity weights are derived from a family expenditure survey covering a wider range of households, but it still excludes certain types of expenditure.

Covariance.   The first PRODUCT MOMENT of two variables about their means. The formula for the calculation of the covariance is

$$\frac{1}{N}\Sigma(X_i - \overline{X})(Y_i - \overline{Y}) \text{ or } \frac{1}{N}\left(\Sigma X_i Y_i - \frac{\Sigma X_i \Sigma Y_i}{N}\right)$$

where X_i and Y_i are corresponding values of each variable, and N is the number of observations.

Example

X	Y	XY
5	1	5
7	2	14
9	3	27
11	4	44
32	10	90

The covariance is

$$\frac{1}{4}\left(90 - \frac{320}{4}\right) = 2.5$$

In the calculation of PRODUCT MOMENT CORRELATION coefficient the covariance (or product moment) is the *numerator*, and the product of the standard deviations of the two variables constitutes the *denominator*.

Covariance, analysis of.   A technique which may be used when some of the explanatory variables are categorized and some are continuous. When variables in continuous form are included in the analysis of variance they are described as covariates and the analysis is relabelled analysis of covariance. The main interest will still normally be on the effect of the categorized variables, and the covariates are included simply to avoid the danger of wrongly ascribing the effect of a covariate to one of the categorized variables.

Coverage.   Formerly the term was used to denote the proportion of the population included in the sample (i.e. the sampling fraction). However, the terms coverage and NON-COVERAGE are now used to describe the extent to which all elements in a population have been included in the SAMPLING FRAME.

The term can also be used to indicate the detail in which material has been collected from the respondents.

Critical region. For the purpose of testing a statistical hypothesis at 1 per cent, 5 per cent or any other level of significance, the total sample space is divided into regions which are mutually exclusive. Sample statistics may either fall into the region of acceptance, which is usually the larger region, or into that of rejection, which is usually the smaller one. The region may consist of two areas. For example, in a two-tail test the region is a combination of two areas, one at each extreme of the curve. The critical region is that of rejection (i.e. of non-acceptance). In the STUDENT'S T-TEST, for example, it is represented by the areas under the curve at one or both extremes beyond a particular CRITICAL VALUE, but this is only one well-known example of a critical region. Generally the term may be applied to the rejection region for any tests of hypothesis. (◆ HYPOTHESIS TESTING.)

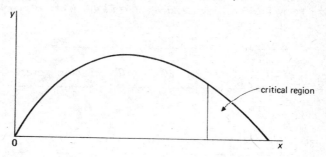

Crossed classification. A two-way or multiway classification of the elements of a population or of a sample according to their scores or levels on a number of variables or factors.

Example
If a company has four breweries and three quality control inspectors who carry out the testing in all four breweries, the resulting data would form a crossed classification and would be presented as follows:

inspector \ brewery	1	2	3	4
1				
2				
3				

In contrast, in a NESTED CLASSIFICATION there would be different inspectors in each of the breweries, and although the data might be presented in the same way the label 'Inspector 1' would have a different meaning in each of the breweries.

Cross-section(al) model.   A model based on, or constructed for, data referring to a particular point in time. Thus the analysis cannot take into account changes over time in the values of the variates. Most social survey data are of this type. In contrast, data in economics frequently take the form of a series of observations over time. With such data, analysis of the kind described in TIME SERIES ANALYSIS can be used.

Cross-tabulation.   A classification in a two-way table of the elements of a population according to their scores or values on two variables is called a cross-tabulation. Each cell contains the number of elements which have common scores or values on each of two variables. The table below presents the cross-tabulation of a group of fifty people by sex and age group:

Sex \ Age group	Under 20	20–39	40–59	60 and over	Total
Male	4	10	8	3	25
Female	3	9	8	5	25
Total	7	19	16	8	50

Such a table presents in summary the information on all fifty people for the two characteristics, age and sex. It is in effect a two-way FREQUENCY DISTRIBUTION. (◆◆ CONTINGENCY TABLE.)

Cumulative error.   An error which does not tend to zero as the number of observations increases. The reverse occurs, for example, in the case of STANDARD ERROR, where error decreases inversely in proportion to the square root of the number of observations or sampling units. If all observations have a bias of a particular magnitude, however, the addition of more observations will not decrease the error. Such an error is called cumulative.

Cumulative frequency.   The sum of all the frequency values of a given variate (x) below and inclusive of a given measure of x for ordinal or interval level variates.

Example
The distribution of earnings in a particular office is shown in the following table:

Salary scale	Frequency (i.e. number of recipients)	Cumulative frequency
£0–£999	10	10
£1000–£1999	60	70 (i.e. 10 + 60)
£2000–£2999	75	145 (i.e. 70 + 75)
£3000–£3999	50	195 (i.e. 145 + 50)
£4000–£4999	40	235 (i.e. 195 + 40)
£5000–£6000	15	250 (i.e. 235 + 15)

Note that the cumulative frequencies are ranked upwards, i.e. £0–£999; £0–£1999; £0–£2999, etc. This is a particular 'interval' case (categorical) of a cumulative frequency distribution. In the more general case, each individual value would be ranked upwards. Cumulative frequency may be expressed in the form of a percentage. (CUMULATIVE FREQUENCY POLYGON, GALTON OGIVE.)

Cumulative frequency curve. OGIVE.

Cumulative frequency polygon. A frequency distribution polygon, which joins together plotted cumulative frequency values of a given variate. If the table illustrating CUMULATIVE FREQUENCY is used, a polygon can be drawn:

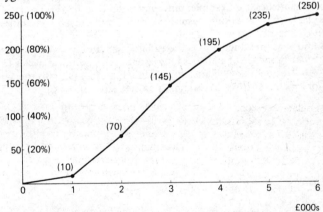

A cumulative frequency curve is a smooth continuous depiction of variate frequencies, but otherwise resembles the cumulative frequency polygon shown above.

Cumulative frequency may be expressed in the form of a percentage. For example, in the cumulative frequency polygon depicted above, the figures in brackets on the right-hand side of the vertical axis are cumulative percentage frequencies, which express cumulative frequencies as percentages of the total number of observations.

The advantage of cumulative percentage frequency curves and ogives is that two or more curves depicting distributions of unequal numbers of observations may be compared using the same diagram. (◆ GALTON OGIVE.)

Curvilinearity.   A relationship between two or more variates which is depicted graphically by any locus other than a straight line. For example, in TIME SERIES ANALYSIS the accelerating rate of increase in prices is called a curvilinear trend and represented by a steepening curve of prices plotted against time. Curvilinearity could represent a logarithmic, quadratic or higher order relationship, or a more complex function.

Cyclical movements/fluctuations.   In *time series analysis* cyclical movements are sequential, repetitive, but not necessarily regular, movements on either side of a given trend line. The object of time series analysis is to discover the extent to which a time sequence of variable values may be predicted by: (1) a trend line; (2) cyclical movements; (3) episodic movements caused by unusual events; and (4) residual random fluctuations. Thus cyclical movements are oscillatory movements superimposed on the trend line in wave-like fashion. In business statistics, for example, they may include both (a) season variations and (b) variations resulting from 'trade cycle' conditions, although the definition is usually reserved for (b) in some statistical texts and seasonal fluctuations are isolated separately.

In other branches of statistics, migratory cycles, political cycles (e.g. between Presidential elections in the United States) fall into the same category. (◆ EPISODIC MOVEMENTS, ERRATIC MOVEMENTS.)

D

Data. The plural of the Latin word 'datum' (=given). The word can mean any information which is 'given' or provided for the solution of a problem. Because more than one unit of information is necessary for solving every problem, the word is plural. Data must be quantified and structured before being usable for the solution of a statistical problem. They are measured in statistical units, which must be: (1) relevant to the enquiry or problem; (2) homogeneous (i.e. of the same type); and (3) stable.

In accounting statistics a measure of homogeneity is often achieved for comparing dissimilar units, for example, the standard hour, a measure of work which a typical individual performs in an hour.

Data are often divided into two categories, PRIMARY and SECONDARY indicating nearness to source or otherwise. For example, data collected by means of experiment, interview, observation, questionnaires or survey are primary, while items of data obtained by the researcher from a previous source or publication are secondary. Often the categories are unclear in practice. (◊ PRIMARY DATA, SECONDARY DATA.)

Data bank. Any comprehensive file of data, whether stored for computing or not. Hence the term can be applied to libraries of secondary data on any given subject.

In the specific computing sense, a data bank is usually a file stored on a direct-access storage device. It can thus be available to different users, often remotely situated from the data bank, and can be accessed by means of terminals, and updated by using a real time system. Airline booking systems often use this kind of data bank.

Data base. A file of data which is designed so that it can satisfy any of a number of different purposes. It is structured in such a way that it may be accessed and updated, but these processes do not modify or limit its content or design. The term is mostly used in a computing context, as DATA BANK.

Decile. One of a series of partition values, known as QUANTILES, which divide the frequency of a group, set, sample or population into equal proportions. As their name implies, deciles divide the total frequency of a set of variate values into ten equal partitions.

Example

The following marks were awarded to nineteen business studies
students for research projects: 40, ㊺ 45, ㊾ 52, ㊲ 55, ㊳ 56,
㊹ 58, ㊺ 60, ㊿ 64, ⑥⑥ 66, ⑥⑧ 69.
As there are nineteen values, the first decile is 44, the second 49, the
third 52, and so on.

As the example indicates, deciles are most frequently used in educa-
tional assessments.

Defective sample.   A sample which is incomplete or results from an
incomplete survey or examination. Two examples of defective samples
are:

1. a sample taken from a telephone directory in an enquiry related to
the study of wealthy people. Although it is valid to assume that most
people whose incomes are above a given level have telephones, some
of the most wealthy people are ex-directory, i.e. their numbers do not
appear in published directories. Such a sample could therefore be
defective;

2. a sample which is incomplete, because some records or completed
questionnaires have been lost.

Deflation of values.   An economic time series measured at current
prices can be converted to a constant price series, giving a better indication of changes in volume by dividing it by an index of the prices of
its components. This technique is referred to as deflating the series.
Some difficulties involved are the choice of base period in terms of
whose prices the deflated series is expressed and the fact that the
relative importance of the components of the series will change as their
relative prices change. This means that a suitable index of prices often
cannot be constructed and a proxy measure must be used such as a
retail or wholesale price index.

Degrees of freedom.   The number of *independent* groups or subcategories into which a sample or population may be divided.

For example, if a population of 300 children is known to consist of
120 males, the two categories male and female may be determined, for,
since it is known that 120 are males, the remaining 180 must be
females. Similarly, if the frequencies of five of six categories into which
a population is divided are known, the sixth may also be determined,
simply by aggregating the frequencies of the five known categories and
deducting their total from that of the population. The frequency of the
sixth category 'depends' on that of the other five categories. In statisti-

cal terminology, there are six categories but only five degrees of freedom.

In the case of a two-way classification, where marginal totals are known, the number of degrees of freedom is $(c - 1)(r - 1)$. To provide a simple example: if it is known that in a school there are 500 boys and 200 girls, 400 successes and 300 failures, it is only necessary to know that 270 of the successes are boys in order to complete all the four categories. We may, by simple deduction, conclude that 130 of the successes are girls $(400 - 270)$, that 230 of the failures are boys $(500 - 270)$ and that 70 of the failures are girls $(200 - 130)$. The knowledge of the frequency of only one subcategory enables us to complete all four. Thus, although there are four subcategories (2×2), there is only one degree of freedom $(2 - 1)(2 - 1)$.

The term degrees of freedom is also used to mean the number of independent comparisons which can be made between the members of a sample or population. Again, the number is $n - 1$, where n is the size of the sample or population. Although the value $(n - 1)$, sometimes indicated by v, is applicable to both these concepts, it is important to realize that the concepts themselves are distinctly different.

Dependent variable. A variable which can be predicted by reference to other variables. The term is used in REGRESSION ANALYSIS to indicate the variable which is likely to have resulted from, or may be predicted by, one or a number of other variables. It is a convention to use the letter Y to indicate the dependent variable and X_i the predictor variables, or regressors, which are used in predicting the dependent variable. Hence in the equation

$$Y = a + bX$$

Y is the dependent variable and X is the predictor, or independent, variable. In this case, for example, Y may be the crop yield per acre and X may be the rainfall in July of the same year. The equation will not in general be deterministic but STOCHASTIC.

The case of linear regression analysis is a specific example chosen to illustrate the meaning of the term, but does not limit its use.

Derived statistics. Statistics obtained by a process of analysis and computation from primary data.

Example
The population of town X is 20 000. During a given year 400 000 books have been issued from the town's libraries. The issue statistic, 20 books per annum per head of population, is a derived statistic.

Descriptive statistics. As STATISTICS has two distinct meanings: (1) the plural of statistic and (2) methods of collecting, presenting and analysing data, so descriptive statistics can mean: (a) statistical data of a descriptive type, or (b) simply calculated measures of data, which are of a descriptive rather than an analytical type. The calculated statistics generally relate to the data on which they are based, rather than to a population from which the data may have been drawn.

Most writers on statistics confine the term to: (i) methods of presenting statistics, using tables, graphs, charts, pictograms etc.; and (ii) some simple methods of calculating measures of central tendency (e.g. mean, median and mode) and measures of dispersion (e.g. standard deviation).

De-seasonalized series.   A time series the observations of which have been adjusted by removing the variation due to seasonal factors (the seasonal variation).

Design effect.   For a particular sample design the design effect (or Deff) is the ratio of the actual sampling variance for that design to the sampling variance of a simple random sample of the same size. Deff is essentially a measure which attempts to summarize the effects of the complexities of the design, particularly clustering and stratification. Deff (or its square root, Deft) is an extremely useful measure in estimating sampling errors for multivariable surveys, although the dependence of Deff on the variable being studied makes it only a rough and ready guide.

Determinant.   The difference between cross-product elements, or the sum of differences between corresponding cross-product elements of a matrix, expressed as a scalar, or matrix. Consider the solution of the following pair of simple simultaneous equations:

$$ax + by = e$$
$$cx + dy = f$$

The matrix of the coefficients of x and y is

$$\begin{pmatrix} a & b \\ c & d \end{pmatrix}$$

The standard algebraic method of solving these equations for the values of x and y would be to multiply all the elements in each line by the coefficients of the other value (i.e. x or y) and to subtract one line from the other. If we require the value of y the first line is multiplied by c and the second line by a, and one line is subtracted from the other, giving

$$y(ad - bc) = (af - ce)$$

so that $y = \dfrac{(af - ce)}{(ad - bc)}$

By a similar process

$$x(ad - bc) = (de - bf)$$

so that

$$x = \frac{(de - bf)}{(ad - bc)}$$

The value $(ad - bc)$ is common to both solutions and is, in the simple illustration given, the determinant of the matrix of coefficients written:

$$\begin{vmatrix} a & b \\ c & d \end{vmatrix}$$

Similarly $(af - ce)$ and $(de - bf)$ are the determinants of the matrices

$$\begin{pmatrix} a & e \\ c & f \end{pmatrix} \quad \text{and} \quad \begin{pmatrix} d & f \\ b & e \end{pmatrix}$$

and are written

$$\begin{vmatrix} a & e \\ c & f \end{vmatrix} \quad \text{and} \quad \begin{vmatrix} d & f \\ b & e \end{vmatrix}$$

where $|D|$ denotes the determinant of the matrix D.

Thus the solutions of x and y may respectively be written as

$$x = \frac{\begin{vmatrix} d & f \\ b & e \end{vmatrix}}{\begin{vmatrix} a & b \\ c & d \end{vmatrix}} \quad \text{and} \quad y = \frac{\begin{vmatrix} a & e \\ c & f \end{vmatrix}}{\begin{vmatrix} a & b \\ c & d \end{vmatrix}}$$

Where matrices contain larger numbers of rows and columns the appropriate value of the determinant depends on the row and column positions of the value for which it is used. Determinants constitute a ready method of carrying out matrix INVERSION.

In some older textbooks the expression 'eliminant' occurs as a synonym for determinant, because it was said to eliminate variables in the solution of simultaneous equations.

Determination, coefficient of. The square of the Pearson product–moment correlation coefficient (ρ) which expresses the proportion of the variance of a dependent variable (y) explained by the independent variable (x), where there is a linear regression of y upon x. In a case of multiple linear regression, where there is multiple correlation between a dependent variable and a set of independent variables, the composite coefficient of determination R^2 is also sometimes known as the index of determination. ($\blacklozenge$ R^2, COEFFICIENT OF ALIENATION and COEFFICIENT OF NON-DETERMINATION.)

Deterministic model.   The antithesis of stochastic model, i.e. a model which contains no random (or chance) elements, and such that all outcomes at any future time are predetermined by the current state, trends and dynamic elements represented by variables in the model itself.

Deviate.   The value (or measure) of a given observation of a variate expressed in terms of its difference from some constant measure of location such as the mean, median or mode. The measure may either be an absolute or standard score. For example, if the mean is 12 and the standard deviation of the variate is 2, and a given observation is 8, the absolute deviate would be 4, but the standard deviate (that is, the deviate expressed in standard score) would be $\frac{4}{2} = 2$.

(◗ ABSOLUTE DIFFERENCE, MEAN DEVIATION, Z-SCORE.)

Dichotomy.   The division of constituents of a sample, set or population into two groups. Although this division is sometimes in terms of a measurable quantity of variable (e.g. over 18 years old or otherwise; under £100 or otherwise), it is usually in terms of attributes (e.g. male or female; correct or defective; old or young, etc.).

Direct effect.   That part of the effect of one variable on another in a PATH MODEL which does not operate through its effect on an intermediate variable in the model. The other part of the TOTAL EFFECT is called the INDIRECT EFFECT.

Discrete series.   This term is used in the literature in two senses: (1) a series which is naturally discrete, i.e. one whose values fall naturally at a number of fixed exact measurements, e.g. a number of fixed integers; and (2) a series which is expressed in discrete form by using class measurements, i.e. CLASS INTERVALS.

Dispersion.   The extent to which observations of a variate are scattered (or different from some centre of moment). Various measures of dispersion have been proposed. The crudest measures of dispersion are (1) RANGE, the difference between two extreme measures of a variate, and (2) INTERQUARTILE RANGE and INTERQUARTILE DEVIATION, because they are based on a limited number of observations. For example, the range of the numbers 1, 49, 50, 51, 52, 53, 1000 is 999, yet the bulk of the values are very close to 51.

MEAN DEVIATION measures dispersion in terms of the absolute difference between each value of a variate and a measure of central location, which may be either the mean, median or mode. The measure is simple to calculate, yet it is not widely used as it does not possess desirable mathematical properties.

STANDARD DEVIATION and VARIANCE are better because: (a) they are intuitively reasonable measures; (b) they have desirable mathematical properties; and (c) the procedure of squaring deviations gives extra weight to extreme observations.

There are other measures of dispersion associated with standard deviation and variance. The COEFFICIENT OF VARIATION, THE COEFFICIENT OF MEAN DEVIATION and the RELVARIANCE may be useful in comparing the dispersions of different frequency distributions.

Finally, associated with Gini is the concept of MEAN DIFFERENCE, which is based on the mean of the absolute differences of all possible pairs of variate values from each other.

Disproportionate allocation. Occurs in stratified sampling whenever different sampling fractions are used in selecting the sample in different strata. The most commonly used form of disproportionate allocation is OPTIMAL ALLOCATION.

Disproportionate stratified sampling. STRATIFICATION denotes the division of the population into subdivisions, known as strata, in each of which sampling is carried out independently. Once the strata have been formed, the investigator has complete freedom to choose the sampling fraction (or proportion of the elements to be selected) in each stratum. DISPROPORTIONATE ALLOCATION (using different sampling fractions in different strata) will be justified on three counts: (1) the more dispersed or variable the elements are in a stratum the higher the sampling fraction; (2) the more expensive it is to obtain information in a stratum the lower the sampling fraction; (3) the more precision we require for the estimates in a stratum the higher the sampling fraction. Disproportionate allocation does, however, involve some risks. Whereas proportionate allocation cannot in general lead to a fall in precision compared to simple random sampling, disproportionate allocation – if based on faulty information about costs or dispersion – can lead to serious losses in precision. The additional costs in the analysis caused by the additional complexity of the sample design should be considered when deciding whether to use proportionate or disproportionate allocation. (♦♦ OPTIMAL ALLOCATION.)

Dissimilarity, coefficient of. Any measure used to describe the difference, distance or dissimilarity between elements, groups or variables. The term is frequently used in CLUSTER ANALYSIS. Distance measures (for example, EUCLIDEAN DISTANCE) are the most commonly used coefficients of dissimilarity.

Dissimilarity matrix (matrix of dissimilarities). A MATRIX in which the COEFFICIENTS OF DISSIMILARITY between each pair of elements in the data set are presented. For example, if there are three elements in the set the distances between them might be

$d(1,2) = 2.7$
$d(1,3) = 1.5$
$d(2,3) = 3.0$

where $d(i,j)$ is the distance between elements i and j.

The distance between each element and itself is defined as zero. Thus the distances can be presented as

$$
\begin{array}{c c c c}
 & 1 & 2 & 3 \\
1 & \begin{bmatrix} 0.0 & 2.7 & 1.5 \\ 2 & 2.7 & 0.0 & 3.0 \\ 3 & 1.5 & 3.0 & 0.0 \end{bmatrix}
\end{array}
$$

The value in the second row of the first column is the distance between elements 1 and 2 and is equal to the value in the first row of the second column, i.e. the matrix is symmetrical, and can be presented without loss of information as

$$
\begin{array}{c c c c}
 & 1 & 2 & 3 \\
1 & \begin{bmatrix} 0.0 & & \\ 2 & 2.7 & 0.0 & \\ 3 & 1.5 & 3.0 & 0.0 \end{bmatrix}
\end{array}
$$

This is the usual form in which the matrix is presented.

Distributed lag.   A term often used in the analysis of time series. It may sometimes be assumed that a particular cause (or series of causes) occurring at one point (or series of points) of time may have effects which are distributed over a number of subsequent time intervals. Thus the effect is not a single period lag, but is distributed over a number of subsequent periods.

Example
Let us assume that normally the reduction of prices in a department store only effects increases of sales during a subsequent week. On one occasion, however, the reductions are advertised in national newspapers and a selling boom occurs over four subsequent consecutive weeks. In this case, the lag is said to be distributed over weeks 1, 2, 3 and 4.

Distributed lag equation.   In econometric models it is often desirable to explain one variable in terms of the values of another in previous time periods. For example, in the permanent income hypothesis, consumption is thought to depend on permanent income which, by one definition, is a geometrically weighted moving average of past periods' incomes.

Distribution curve. The graph of a distribution function, showing frequency values plotted against an ordinate and variate values against an abscissa. This term is popularly used to denote a CUMULATIVE FREQUENCY CURVE.

Distribution-free methods. Methods of testing given hypotheses, which do not depend on assuming a particular parametric frequency distribution, but which are applicable to all frequency distributions. It is possible to conduct tests involving ranked observations from elements of a sample, whose results would be valid quite irrespective of the parameters of the population from which the sample is drawn. Such methods of testing are distribution-free.

The term NON-PARAMETRIC METHODS is sometimes used interchangeably with distribution-free methods.

Distribution function. Although the general meaning of this term may include any frequency distribution, where a given variate x, for example, may have a frequency 5 at x_1, 10 at x_2, 12 at x_3, etc., the term is often specifically used to mean the total (or cumulative) frequency of components of a series with variate values less than or equal to x_i.

In the above example, the distribution function $F(x)$ would take the frequency value 5 at x_1, 15 (i.e. 5 + 10) at x_2 and 27 (i.e. 15 + 12) at x_3, etc. Thus the term is associated with CUMULATIVE FREQUENCY, CUMULATIVE FREQUENCY CURVE.

Domain of study. A domain is a part of the population in which the investigator has a special interest and for which separate estimates are desired. STRATIFICATION, where possible, may be used to separate the domains of study so that the desired precision can be built into the survey design. Although the term should be used to denote subdivisions of the population for which estimates with known precision are planned, it is sometimes used to specify subdivisions for which separate estimates are to be calculated but for which the requisite prior information for the design is not available. The term subclass would be more appropriate in these cases.

Double sampling (multiple sampling). An extension of ACCEPTANCE SAMPLING. If, for example, the acceptance number of defectives for a particular sample is 2, and two defectives occur, there is a problem whether the batch from which the sample is taken should be accepted or rejected. If they are clerical records, for example, complete acceptance (such as if 0 or 1 defective record had occurred) would involve considerable risk, while complete rejection of the whole batch (such as if more than two defective records had occurred) would involve the uneconomic cost of checking every record in the batch. A compromise between the two procedures is that, if there are two defectives in the first sample of 200, a second sample should be drawn from the batch, using a sample size of 100. If there are any defectives in the second sample reject the whole batch, but if there are none accept it. The effect of having no defects in the second sample is to make the proportion of defects fall below the acceptance number for the extended sample (2 *for 200* + 0 *for 100* = 2 *for 300*). The batch thus becomes acceptable. (◆◆ CONTINUOUS SAMPLING (sequential sampling) and TWO-PHASE SAMPLING.)

Down cross/up cross.   A point in time series analysis where the locus of time series points crosses the trend line downwards so that the deviations of the actual time series values from the trend line become negative. Conversely the up cross is the point where the time series line crosses the trend line upwards, so that the deviations of the time series line from the trend line change from negative to positive.

Dual theorem (dual problem).   A term used in linear programming to indicate a reverse problem to its counterpart, often styled the primal problem.

Example
A firm uses two restricted inputs, paper and labour, to produce newsprint and books. Every batch of newsprint sold adds £4 to the firm's profits and every batch of books (100) sold adds £5 to the firm's profits. There are only 200 units of paper in a given period of which 5 are needed for book production and 3 for newsprint, while 600 hours of labour are available in a given week, of which 7 hours are required for a batch of newsprint and 4 hours for a batch of books. The primal model would thus be formulated.

$$
\begin{array}{lll}
\text{Objective:} & \text{Maximize} & 4x_1 + 5x_2 \\
\text{Constraints:} & \text{Subject to} & 3x_1 + 5x_2 \leqslant 200 \\
& & 7x_1 + 4x_2 \leqslant 600 \\
& & x_1, x_2 \geqslant 0
\end{array}
$$

Note that x_1 and x_2 are batches of newsprint and books to be produced.

Thus the objective is to maximize profit. However, the problem can be restated as a cost-minimization problem, which appears similar to the original problem but turned on its side;

$$
\begin{array}{lll}
\text{Objective:} & \text{Minimize} & 200a_1 + 600a_2 \\
\text{Constraints:} & \text{Subject to} & 3a_1 + 7a_2 \geqslant 4 \\
& & 5a_1 + 4a_2 \geqslant 5 \\
& & a_1, a_2 \geqslant 0
\end{array}
$$

The solution of the primal problem provides the number of batches of newsprint and books which ought optimally to be produced in a given period, whereas the solution of the dual problem provides the shadow-costs or opportunity costs of production; these will be identical with the likely contribution (a) per unit, given that the optimum quantities of each output are being produced.

Dummy variable. A variable in a regression or other equation which takes the value 0 or 1 and usually represents an attribute or categorization. Thus it is an artificial variable expressing qualitative characteristics. For example, in multivariate regression analysis, some of the REGRESSORS may be attributes (e.g. sex) and take values 0 or 1.

In some cases dummy variables may take more than two values, e.g. the values -1, 0 and $+1$. In a more rigorous sense the constant may itself be regarded as a dummy variable ($= 1$), so that a multiple regression model can assume the form (with $\beta_\alpha x_0 = \beta_0 1$)

$$y = \sum_{i=0}^{n} \beta_i x_i + \varepsilon$$

Durbin–Watson statistic. One of the assumptions of the ordinary least-squares method of estimating regression coefficients is that the disturbance (stochastic) terms are uncorrelated with one another. If the data used are recorded sequentially through time (time series data) then it often happens that these terms in the regression model are autocorrelated. A test for autocorrelation has been devised by Durbin and Watson. If $e_1, e_2, \ldots e_n$ are the unexplained residuals from the fitted regression equation the Durbin–Watson statistic is

$$d = \frac{\sum_{t=2}^{n} (e_t - e_{t-1})^2}{\sum_{t=1}^{n} e_t^2}$$

and its value lies between 0 and 4. If the model stochastic terms are uncorrelated (at lag 1) d will take values near 2; smaller values indicate positive autocorrelation and higher values negative. A test of the hypothesis that the autocorrelation is zero can be made by comparing the value of d with a table of significance points. It turns out that the true significance points of the d statistic depend on the independent variables in the regression, but upper and lower limits for the significance points have been found that do not depend on the values of the independent variables. For values of d between 0 and 2, if d is less than the lower limit then the hypothesis can be rejected at the appropriate level. If it is above the upper limit the hypothesis cannot be rejected; and in between the test is inconclusive. If negative autocorrelation is likely to be present, use $d^* = 4 - d$ and the same upper and lower significance limits apply.

Dynamic model. A model in which time factors are involved, and therefore sometimes termed in econometrics a multitemporal as distinct from a unitemporal model. Three main uses of the term are illustrative:

1. In computer simulation, a dynamic model is one which is designed to show the effect of the passage of time on a number of variables. If, for example, a trade cycle or discount factor or annual loop is used in the programme, the model is a dynamic one.

2. In time series analysis, a model may express conditions at time $t + 1$ in terms of conditions at time t. This kind of model may be dynamic if it can be applied to subsequent period.

3. Ideally in econometric usage a dynamic model has at least one variable whose values fluctuate over time, and at least one equation which represents a dependent vaiable as a function of time.

E

e. ◆ EXPONENTIAL.

Econometric model.   An econometric model is a set of one or more structural equations whose form (i.e. which variables depend on which) is derived from economic theory and in which any unknown parameters are estimated statistically. There are normally as many structural equations in the model as there are endogenous variables that the model seeks to explain. Each equation relates one endogenous variable to other endogenous and exogenous variables. A simple example is the two-equation model of the market for an industrial product consisting of supply and demand questions. The endogenous variables in this case would be price and quantity, and exogenous factors might include advertising expenditure (affecting demand) and research and development (r & d) expenditure (affecting supply). Thus we might have

Quantity = $a + b \times$ price $+ c \times$ (r & d) $+ u_1 \rightarrow$ supply equation
Quantity = $d + e \times$ price $+ f \times$ advertising $+ u_2 \rightarrow$ demand equation

These can be written as:

$$y = a + bx_1 + cx_2 + u_1$$
$$y = d + ex_1 + fx_3 + u_2$$

where x_1 represents price, x_2 represents research and development, x_3 represents advertising and where a, b, c, d, e, f are coefficients to be estimated and u_1, u_2 are unexplained residual terms. Given values for the exogenous variables which can be fixed by the firms, we can solve the equations simultaneously to find values for price and quantity to be sold.

More generally it is often convenient to use a shorthand matrix notation to express the equations of a model. Let y be the $p \times 1$ vector of endogenous variables, x be a $q \times 1$ vector of exogenous variables, B be a $p \times p$ matrix the elements in the jth row of which are the coefficients of the endogenous variables in the jth equation, Γ be a $p \times q$ matrix the elements in the jth row of which are the coefficients of the exogenous variables in the jth equation, and u be a $p \times 1$ vector whose elements are the unexplained residuals in the equations. Then the system can be written as

$$By + \Gamma x = u$$

By convention we set one non-zero element in each row of B (often the diagonal element) equal to 1, which we are free to do since multiplying an equation through by a constant does not affect its validity as a description of the economic mechanism operating.

A particularly important problem when dealing with simultaneous equation models is that of estimating the structural coefficients. Ordinary least-squares regression cannot be applied to equations of this nature since the residual in each equation is not independent of all the explanatory variables whose coefficients are being estimated.

Equations such as this, where the effect of a change in the determining or 'independent' variable is felt progressively with the lapse of time, are called distributed lag models.

Econometrics. Definitions of broad areas of knowledge and study are necessarily somewhat loose as one subject area frequently overlaps into several others. This is certainly true of econometrics, which, broadly speaking, is concerned with quantifying economic theories and relationships. Thus it plays a role to economics that is in many ways similar to that of the engineer in relation to the pure scientist.

There appear to be three broad strands to econometrics. It is necessary first to express economic postulates and theories in a form that is susceptible to empirical estimation and verification. This involves expressing theories as equations or other mathematical constructs, and in doing this the mathematics sometimes assumes a momentum of its own leading one into mathematical economics. The second branch consists of the statistical theory necessary to solve the estimation and other inference problems that arise when one seeks to quantify economic thought. This consists mainly of regression theory and its application to systems of simultaneous equations in several endogenous variables, although other branches of statistics – for example, time series analysis, multivariate analysis and distribution theory – are also involved. Much of this theory has come to be regarded as part of econometrics rather than statistics because it was developed in response to the needs of the econometrician and often by him. The third strand of econometrics consists of the body of empirical work that has been built up through measuring economic variables and estimating relations between them using the techniques of the first two branches. It is sometimes referred to as economic statistics, or applied econometrics, and includes the analysis and use of econometric models that have reached the stage of having had coefficients estimated.

Some authors prefer to delimit the scope of the subject much more severely than this, concentrating mainly on the second strand but sometimes willing to admit parts of the third. On the other hand, most econometric journals take a fairly broad view. Within this subject area there appears to be a consensus that the following topics should be covered in any course or text in econometrics: general linear regression theory, simultaneous systems, identification, methods of estimation and properties of estimators, autocorrelation, lag structures and dynamic systems.

Editing. Editing of completed questionnaires is used to identify, and if possible eliminate, some of the errors which are present in the questionnaires. The editing should be carried out as soon as possible – (1) by the interviewer, if the responses are obtained through an interview (and subsequently by the office staff), or (2) by the office staff if the questionnaire is self-administered. The three principal objectives are to check for completeness, for accuracy and for uniformity in the interpretation of the questions. Nowadays much of the editing is done by computer, and many powerful computer programs are now available to perform simple and complex CONSISTENCY CHECKS. However, the initial editing must still be done by hand (or eye) since the computer cannot be used until the responses are coded.

Effective range. A range over which values are distributed exclusive of exceptional values, and can therefore be taken as a more realistic measure of difference between upper and lower values than the actual range itself. For example, in the series 1, 79, 80, 81, 82, 84, 90, 600, the first and last values are extreme values, and the effective range of values $(90 - 79 = 11)$ may be a better measure of difference between upper and lower values than the range $(600 - 1 = 599)$.

Efficiency. A desirable property of estimators or testing procedures in statistical inference. (◗ EFFICIENT ESTIMATORS, EFFICIENCY OF AN ESTIMATOR, EFFICIENCY OF TESTS.)

Efficiency of an estimator. For an estimator which is asymptotically (for large samples) normally distributed and consistent, efficiency is defined as the ratio of sample sizes required to make its sampling variance equal to that of the EFFICIENT ESTIMATOR as the sample size increases. Since sampling variances will depend on sample sizes this is just a way of assessing just how good a particular estimator is relative to the best one available for large samples.

Efficiency of tests. For a particular null hypothesis and a simple alternative, a given sample size, and a fixed size of test (i.e. specified probability of type I error), then the best test procedure will be that with the smallest type II error (or largest power). This is referred to as the efficient test for this situation. Other test procedures can be compared to the efficient test by examining their relative efficiencies. These are defined as the ratio of the sample size one test requires to the size specified for the efficient test to make the powers equal. Often an inefficient test may be preferable if its relative efficiency is high because it is conceptually simpler and perhaps easier to calculate and manipulate.

Efficient estimators. Most common estimators have asymptotically (for large sample sizes) normal sampling distributions. Consistent estimators will also be asymptotically unbiased. Among such estimators the one with the smallest variance in large samples is called an efficient estimator. This acts as a standard against which the usefulness of other estimators may be judged.

Element. A member of a group, set or collection of items which cannot be further subdivided into constituents that could be regarded as members of a collection. Hence the mathematical and statistical usage of the word is analogous to its usage in the physical sciences. Two common usages of the word are:
1. in MATRIX ALGEBRA where, for example, the sixteen constituent members of a 4×4 matrix are elements since they cannot be further subdivided in the same way as the matrix;
2. in computing, where the constituent elements of a record are words, and the elements of each computer 'word' are described as bits (BINARY digits).

Element sampling. A sampling scheme in which the sampling unit contains only one element. This is in contrast to CLUSTER SAMPLING, in which groups of elements are selected together.

Elementary unit/element. The smallest unit yielding information. In many cases the element in sample surveys is the individual person. In some cases the element may be a household, a firm or even a county. The element may, however, be different from the UNIT OF ANALYSIS.

Endogenous. In multivariate analysis the variables can be divided into two classes – endogenous and EXOGENOUS. Those endogenous to the system are those which are influenced by other variables within the system. The term is sometimes used as a synonym for DEPENDENT, e.g. in a regression equation. A variable may be endogenous in one model and exogenous in another. For example, rainfall might be endogenous to a meteorological model dealing with climate but exogenous to a model of demand for sports equipment.

Endogenous variable. This is a variable in an econometric model that is explained by the model. Thus it plays the same role as the dependent variable does in a regression equation. The model itself is a simplified representation of a complicated system in which many endogenous variables are being simultaneously determined. The driving force of the system is provided by the exogenous variables – those whose values are fixed without reference to the system under discussion. The distinction between exogenous and endogenous variables in practice is often an artificial one, and is dictated by the practical limits on the size of the model that one is able to construct and the use that is to be made of it.

Episodic movements. In a time series of observations, these are the atypical movements caused by unpredictable isolated or irregular events which occur from time to time. They include strikes, wars, epidemics, earthquakes, major meteorological events. They must be sufficiently severe to cause an identifiable departure in the time series from its normal course.

Epsem sampling. The word 'epsem' denotes an Equal Probability of Selection Method. Epsem sampling therefore includes any sampling scheme in which every element in the population has an equal probability (chance) of selection. The big advantage of epsem is that it leads to SELF-WEIGHTING SAMPLES, in which there is no need to give different weights to different elements in order to obtain good estimates of the population parameters.

Erratic movements. In a time series these are irregular movements as opposed to trend or cyclical components. They include the ordinary random fluctuations that continually occur but exclude longer departures that can be traced to specific and isolated incidents such as wars and strikes; these latter are referred to as episodic movements. When analysing the time series we often wish to smooth out the irregular or random terms to get a picture of the underlying series, on the grounds that such erratic movements are transitory and will not continue to affect the series in future periods.

Error. Although the word often means a mistake in the general sense, it has a number of other meanings in statistics. It may mean an approximation difference. For example, when 950 is described as approximately 1000 ABSOLUTE error is 50 and relative error is 5 per cent. Alternatively it may mean an error of observation, that which arises from imperfect means of observation whether mechanical or human; or an error of reference, which results from wrongly attributing to one cause rather than another; or a copying error, which results from wrongly transcribing data.

The word is often specifically used to indicate the difference between an actual value and its expected value, or estimate. For example, a regression equation may be used to estimate the price of a good on a given day as £500. If the actual price is £495, the error (which could perhaps be attributable to a 'chance' effect) is £5. The term RESIDUAL is more appropriate in this case.

Error of observation.   An error which arises from faults in the method of observing a quantity. Errors of observation may be either due to faulty instrumentation or to human error. (◆◆ MEASUREMENT ERROR, RESPONSE ERROR.)

Error sum of squares.   The sum of squares of RESIDUALS between observations and estimates using a particular multiple regression model. In ANALYSIS OF VARIANCE it is assumed that observations result (1) from the effect of a number of REGRESSORS whose VARIANCES can be estimated and (2) from a random component, i.e.

$$y = \sum_{i=1}^{p} \beta_i x_i + \varepsilon$$

where a linear model is used.

If one makes estimates from a model corresponding to each of the observations, the difference between each observation and its corresponding estimate is a residual, assumed to have resulted partly from the stochastic component in the model and partly from errors of estimation of the parameters in the model. Thus if the model is $y = X\beta + u$ and β is estimated by $\hat{\beta}$, $\hat{y} = X\hat{\beta}$ and therefore:

$$(y - \hat{y}) = X\beta + u - X\hat{\beta}$$
$$= X(\beta - \hat{\beta}) + u$$

The term RESIDUAL sum of squares is in some ways more appropriate than error sum of squares.

Essential survey conditions.   In setting up a social survey the researcher can specify to a considerable degree some of the basic characteristics of the survey and therefore the conditions under which the information is collected. Among the important elements involved are the method of hiring and training interviewers and the procedures used for recording and analysing the responses. Any shortcomings in these essential conditions may lead to serious biases in the data collected.

Estimate.   In estimating some characteristic of the population from a sample, the value obtained from a particular sample is an estimate. In short, it is the particular value obtained for an ESTIMATOR in a given set of circumstances.

Estimation.   Inference about the numerical values of a population from a sample or from incomplete data. To state, for example, by examining a sample, that the mean height of persons in a group is estimated as 1.3 metres, is called POINT ESTIMATION. More often, an interval is calculated, within which the parameter values are likely to lie, for example, between 1 and 2. This is known as an INTERVAL ESTIMATE.

Sometimes the word is used in respect of REGRESSION ANALYSIS or the prediction of a series of values of a regressand, using the values of the regressors and a regression model.

Estimator.   A method or rule for estimating from a sample a characteristic of the population being studied. For instance, to estimate the proportion of the population members who have blue eyes the proportion of the sample who have blue eyes may be used as an estimator. The value of this sample proportion obtained from a particular sample is called an estimate.

Euclidean distance.   The Euclidean distance metric is the most commonly used distance measure in CLUSTER ANALYSIS, and also the most familiar. the distance between two points i and j, d_{ij}, is defined as

$$d_{ij} = \left[\sum_{k=1}^{p} (x_{ik} - x_{jk})^2 \right]^{1/2}$$

A simple example may serve to illustrate its application. In the two-dimensional case below, the distance between points 1 and 2 is given by the length of the straight line connecting the points, i.e. the distance d_{12}.

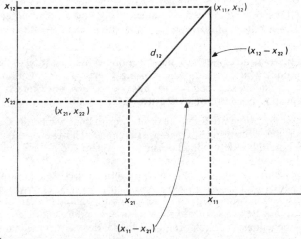

Here

$$d_{12} = [(x_{11} - x_{21})^2 + (x_{12} - x_{22})^2]^{1/2}$$

(◆◆ CITY BLOCK METRIC; DISSIMILARITY, MEASURES OF.)

Event.   In probability this term is employed to mean any one of a set of probable outcomes or happenings, which in simple illustrations of probability are always mutually exclusive. For example, if a die is cast, the outcome '6' is an event, which excludes the outcomes '1', '2', '3', '4' and '5'.

In more complex illustrations of probability, where events are not necessarily mutually exclusive, as illustrated in VENN DIAGRAMS, the illustrative space corresponding to a given event is sometimes known as event space, although in most texts this term is identical with SAMPLE SPACE.

ex ante.   'From beforehand.' The term is used in the literature in association with models which may postulate beforehand a particular state or set of events. For example, probabilistic models and probability distributions, time series and other econometric models are of this type. In accounting statistics the term is sometimes employed for unmodified budget estimates, based on statistics available when a budget is compiled.

ex post.   'From afterwards.' The term generally means a retrospective view of an event or set of events. Actual observed results may be compared with those predicted by a given model, and the resultant comparison is ex post. Any modification of the model or hypothesis is also termed ex post. In financial statistics the term is sometimes employed for modified budget estimates, taking into account exceptional events which have occurred since the original EX ANTE estimates were made.

Exhaustive classification.   A classification or categorization of the variate values which includes all possible values. If, in addition, each possible value can belong to one and only one category the classification is said to be exhaustive and mutually exclusive.

Exogenous.   In multivariate analysis the variables can be divided into two classes – ENDOGENOUS and exogenous. Those exogenous to a system are those determined outside the system and not influenced by other variables in the system (or such determination is not of interest in the model). The term is sometimes used as a synonym for INDEPENDENT e.g. in a regression equation. A variable may be endogenous in one model and exogenous in another. For example, rainfall may be exogenous to a model of demand for sports equipment but endogenous to a meteorological model dealing with climate.

Exogenous variable.   In an econometric model an exogenous variable is one that is determined outside the model and is not itself jointly dependent with the endogenous variables explained by the model. A simple example would be the level of public spending in a model of a national economy. This is usually thought to be fixed exogenously by the Government, at least in the short term, and not to be jointly dependent with the endogenous variables such as consumption, income, investment, etc.

Expected value/expectation. The mean value of an ESTIMATOR or function in repeated sampling is its expected value. Thus, if an estimator could take the values 3, 7, 12, 2, 15, 9 each with probability 1/6, the expected value of the estimator is 1/6[3 + 7 + 12 + 2 + 15 + 9] = 48/6 = 8. Thus the expected value need not be a value which itself occurs at all. For example, with a fair die (each side having equal probability of occurring in a single toss) the expected value is 1/6[1 + 2 + 3 + 4 + 5 + 6] = 3½, which can never occur. The symbols $E[\ ]$, $\varepsilon[\ ]$ and EV[ ] are commonly used to denote expected value.

Experiment. From the Latin 'experimentum' (derived from 'experiri', to 'try thoroughly'). The term is used for any method of obtaining primary data or of testing an hypothesis, where the conditions of the investigation or test are controlled.

Although experiments are most commonly employed in the physical sciences, their usage is more extensive. For example, a temporary traffic roundabout may be constructed to test the effectiveness of a roundabout before permanent installation. In this case, not only is physical effectiveness being tested, but also social usage.

Explanatory variable(s). If a relationship between two variables x and y is interpreted in a causal way, such that changes in y are said to be caused or explained by changes in x, then x is called an explanatory variable. The term is used loosely to be synonymous with terms such as INDEPENDENT VARIABLE, predictor variable and REGRESSOR.

Exponent. The power to which a number is raised. Hence if a number is squared the exponent is 2. In the case $10^{8.8}$ the exponent is 8.8.

Exponential (*e*).   The sum of the series

$$e = \frac{1}{0} + \frac{1}{1!} + \frac{1}{2!} + \frac{1}{3!} + \frac{1}{4!} + \frac{1}{5!} + \frac{1}{6!} + \frac{1}{7!} + \frac{1}{8!} + \ldots \frac{1}{\infty!}$$

This is

$$e = 1 + 1 + 0.5 + 0.167 + 0.04167 + 0.00833 + \ldots \text{ etc.}$$

and approximates to 2.7183.

The value is important in a number of statistical contexts, e.g. (1) the calculation of the POISSON DISTRIBUTION for any given value of λ; (2) the calculation of the area under a normal curve. The limit '*e*' is the base of Napieran (or natural) logarithms.

Exponential curve. A curve representing the continuous or limiting form of an exponential series, or employing continuously an exponential equation such as

$$y = ae^{bx}$$

where y is the dependent variable, x the independent variable, and a and b are constants. Curves showing:
1. increases of value at compound interest rates over time,
2. increases of prices if inflation continues at a fixed rate,
3. depreciation at *reducing balance method* (in accounting), and
4. the rate of forgetting memorized material (the Ebbinghaus Curve),
are all examples of exponential curves where x in the exponent (bx) is a time sequence.

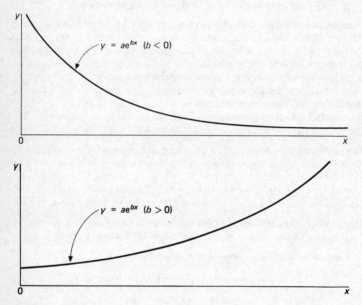

Exponential curves

Exponential distribution. A distribution where

$$dF = \frac{1}{\sigma} \exp \left(- \frac{x - m}{\sigma} \right) \ dx, \ m \leqslant x \leqslant \infty$$

and where σ is not only the standard deviation of the distribution but is the distance of the mean from the commencement of the distribution. (◆ EXPONENTIAL CURVE.)

Exponential smoothing. A computationally simple way of removing random fluctuations from a time series. In its simplest form, given a time series of observations $x_1, x_2, \ldots$, the exponentially smoothed value S_t at time t is given by

$$S_t = a\, S_{t-1} + (1 - a)\, x_t$$

In this formula a is the smoothing constant, and by varying the choice of this constant it is possible to vary the rate of response of the smoothed series to a change in the level of the series. By repeated substitution it is easy to show that

$$S_t = (1 - a) \sum_{j=0}^{\infty} a^j x_{t-j}$$

is a moving average with geometrically declining weights attached to past values of the series. In this form it is suitable for estimating the means (and predicting them) for a series with no trend or cyclical movements. A major advantage of the method is that it involves very little storage of information from one period to the next. In fact, only the smoothing constant and the previous smoothed value need be remembered. More complicated forms such as double exponential smoothing can be used when the series contains a trend.

Extremal quotient. The relative difference between the largest and smallest values in a sample, as the RANGE is the ABSOLUTE DIFFERENCE.

Example
In a set of observations the smallest value is 20 and the largest value is 600. The difference between these extreme measurements may be computed either: (1) by absolute measurement, using range (i.e. $600 - 20 = 580$); or (2) by relative measurement, using the extremal quotient, i.e. $\dfrac{600}{20} = 30$.

F

F-test. A test, devised by Snedecor, for comparing two separate estimates of the population variance or for comparing estimates of the variances of two populations. It is the basic test used in ANALYSIS OF VARIANCE. A simpler example is given below.

The basic procedure is to take the ratio of the two variance estimates; this ratio is denoted by F. The ratio F has a known sampling distribution provided the two estimates of variance are actually independent of each other, and consequently a fairly simple test can be made.

The ratio will show how plausible it would be to suppose that the two samples came from populations with the same variance. If it is higher than the critical value at a given SIGNIFICANCE LEVEL the ratio of the two variances is said to be significant at that level.

The test was named 'F' by Snedecor, in honour of R. A. Fisher, because it depends fundamentally on Fisher's z distribution.

Example
Two samples have sums of squares of 79 239 and 16 422. The first sample contains 11 units and the second sample consists of 10 units. The DEGREES OF FREEDOM (DF) are 10 and 9 respectively. Are the variances of the samples significantly different?

Procedure
A variance table is compiled, estimating the population variance, by dividing the sample sums of squares by the appropriate number of degrees of freedom. The ratio of variances to each other is the F-value.

Sample sum of squares	DF	Estimated population variance	F-value
79239	10	7923.9	$\left(\dfrac{7923.9}{1824.7}\right) = 4.342$
16422	9	1824.7	

At the 5 per cent significance level, when the samples have 10 and 9 degrees of freedom respectively, the critical F-ratio value is 3.19, while at 1 per cent significance level the relevant F-value is 5.26. In this case, therefore, the difference between variances is significant at 5 per cent significance level but not at 1 per cent significance level.

Face validity. The most superficial examination of the VALIDITY of an attitude scale involves merely checking that all the items in the scale are dealing with some aspect of the attitude continuum under study. In measuring workers' attitude to their supervisors, for instance, we would check that each item was concerned with some aspect of the relationship between workers and supervisors. If this is so, the scale has face validity. This is the minimal check necessary in constructing an attitude scale.

Factor. This term is used in many contexts:
1. a variable by which we stratify – a STRATIFICATION FACTOR;
2. a quantity or variable being tested in the analysis of variance;
3. in multivariate analysis a combination, usually linear, of the variables which is considered to represent an important underlying dimension;
4. in algebraic expressions, in the usual mathematical sense.

Factor reversal test. One of the properties sometimes required of an 'ideal' index number is that, given a set of data from which price and quantity indices can be calculated, the product of the price index and the quantity index should give the change in value that has taken place. In other words, if the index satisfies this test then it is possible to split the change in value (of purchases, say) exactly into a price component and a quantity component by means of a single index. The roles of the price factor and the quantity factor in the calculation of value should be reversible. This property is not satisfied by either the Laspeyres or the Paasche index numbers. For example, consider the Laspeyres index

$$\underset{\text{(Price index)}}{\frac{\Sigma q_0 p_1}{\Sigma q_0 p_0}} \quad \times \quad \underset{\text{(Quantity index)}}{\frac{\Sigma p_0 q_1}{\Sigma p_0 q_0}} \quad \neq \quad \underset{\text{(Value index)}}{\frac{\Sigma p_1 q_1}{\Sigma p_0 q_0}}$$

Fisher's 'ideal' index, however, does satisfy the test.

Factorial. The factorial of a number is the product of a series of digits from 1 up to and including the number. It is indicated by placing an exclamation mark (!) after the number.
   Thus

   2! (2 factorlal) is $2 \times 1 = 2$

while

   7! (7 factorial) is $7 \times 6 \times 5 \times 4 \times 3 \times 2 \times 1 = 5040$

Zero factorial (written 0!) is defined to be equal to 1, i.e. $0! = 1$.
The concept is important in such cases as:
1. calculating probability in sampling without replacement;
2. calculating expected values in the POISSON DISTRIBUTION;
3. computing probability in FISHER'S EXACT PROBABILITY TEST.
The term should not be confused with FACTOR.

Filter questions. It is undesirable that questions should assume any knowledge or behaviour on the part of the respondent. For example, the question, 'Did you enjoy watching the Prime Minister on television last night?' does not leave a reasonable direct response for those who did not watch the programme. Such questions should be preceded by filter questions designed to find out whether the main question is applicable to the respondent. In this case the filter questions might be (1) 'Did you watch television last night?' and, if Yes, (2) 'Did you watch the Prime Minister on television last night?'. Similarly as a filter to the question 'How did you vote at the last election?' we should use 'Did you vote at the last election?'.

Finite multiplier/finite population multiplier. A term which is synonymous with FINITE POPULATION CORRECTION and is not now widely used although it was at one time considered a preferable term.

Finite population correction. Most statistical theory is based on the assumption of UNRESTRICTED RANDOM SAMPLING, in which case the variance of the sample statistics is independent of the population size N. When the sampling is done without replacement the sampling variance depends also on N. Thus the variance of the sample mean $\bar{x}$ can be written as:

$$\text{Var}(\bar{x}) = \left(1 - \frac{n}{N}\right) \cdot \frac{S^2}{n}$$

where S^2 is the population variance $[1/(N-1) \sum^{N} (x_i - \mu)^2]$ and n is the sample size. The factor $(1 - n/N)$ is called the finite population correction. This usage was considered objectionable at one time since the factor is not strictly speaking a correction but is an intrinsic part of the formula. However, it is now in common use. The term FINITE POPULATION MULTIPLIER, or FINITE MULTIPLIER, may also be used.

First moment, second moment, moment. A moment is the mean value of the power of a variate. In a univariate distribution, the first moment is the arithmetic mean of that distribution, the second moment is the mean of its squares, the third moment the mean of its cubes and so on.

In a bivariate case, the term product moment is employed to denote the mean of the products of corresponding pairs of values of the two variates.

The meaning of the term may be extended to and employed in multivariate distributions.

Fisher's exact probability test, Fisher's test. A method of testing association in 2×2 contingency tables where the total number of observations is less than fifty, and where, in consequence, Yates modification of the CHI-SQUARED TEST is unreliable.

The value of P is obtained by using the formula

$$P = \frac{(a+b)!(c+d)!(a+c)!(b+d)!}{n!a!b!c!d!}$$

where a, b, c and d are the cell frequencies and n is the total number of observations (i.e. $n = a + b + c + d$). The NULL HYPOTHESIS is that of non-association between row and column categories. If the value of P falls below 5 per cent or 1 per cent, the null hypothesis is rejected at the 5 per cent or 1 per cent levels respectively.

Example

In an examination the results for two schools are as follows:

	A	B	total
pass	11	3	14
failure	6	15	21
total	17	18	35

We assume that the null hypothesis, that there is no significant difference between the performance of candidates from schools A and B is correct, and test the hypothesis by calculating P, the probability that the observed results would occur if the null hypothesis were true. In this case:

$$P = \frac{14! \times 21! \times 17! \times 18!}{35! \times 11! \times 3! \times 6! \times 15!}$$
$$= 0.5\%$$

The null hypothesis can thus be rejected at the 1 per cent level. There is evidence of significant association between the schools (A and B) and the category of result (pass or fail).

Fisher's ideal index. Three properties that it is sometimes argued an 'ideal' index number should possess are that it should satisfy the circular test, the factor reversal test and the time reversal test. In fact no index number satisfies all of these but Fisher has suggested an index that satisfies two of them. It is the geometric mean of the Laspeyre and

the Paasche indices. The price index for time period 1 based on period
$0 = 100$ is

$$F_1^{(0)} = 100 \times \sqrt{\left\{ \left[\frac{\sum\limits_{i=1}^{n} p_i^{(1)} q_i^{(0)}}{\sum\limits_{i=1}^{n} p_i^{(0)} q_i^{(0)}} \right] \left[\frac{\sum\limits_{i=1}^{n} p_i^{(1)} q_i^{(1)}}{\sum\limits_{i=1}^{n} p_i^{(0)} q_i^{(1)}} \right] \right\}}$$

The only condition that Fisher's index fails to meet is the circular
test, i.e.

$$\frac{F_1^{(0)}}{100} \times \frac{F_2^{(1)}}{100} \times \frac{F_0^{(2)}}{100} \neq 1$$

Despite the fact that it has two of the three 'ideal' properties it is rarely
used in practice because of the relative complexity of the calculations
involved.

Follow-up.   A further attempt to obtain information from an individual in a survey when the first attempt has been unsuccessful. The
term is often used specifically when the individual has moved from the
address given in the original listing. The term CALL-BACK is more
common when the later attempts involve calling at the same address.

Frame (for sample selection).   A term used to denote a list, map or
other specification of a population which is used as the basis for selecting a sample using a particular sampling scheme. (♦ SAMPLING FRAME.)

Frequency.   The number of occurrences of:
1. a particular event; or
2. a particular class or interval group; or
3. particular values of a variate.
   For example, if in a degree examination there are three distinction
results, twelve passes with credit, thirty ordinary passes and two failures, then the frequency of distinctions in the examination is three, etc.

Frequency curve.   A graphical depiction of a FREQUENCY DISTRIBU-
TION. Frequency curves are depictions of continuous frequency distributions, as distinguished from FREQUENCY POLYGONS which are used
to illustrate discrete frequency distributions. Thus frequency curves are
used where class intervals become very small and numbers of observations are very large. The ordinate axis is used to measure frequency
values and the abscissa is employed to measure variate values.

Example

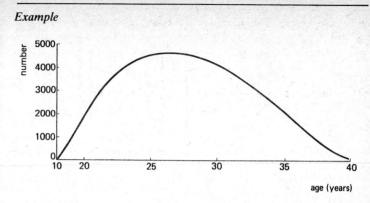

Numbers and ages of people taking a given professional examination

Frequency distribution. A depiction in either graphical or tabular form, indicating the manner in which the frequencies of constituents of a sample or population are distributed. Frequency distribution tables show how a sample or population is divided from the standpoint of frequency into interval classes of a variable or into attribute categories, and BAR CHARTS and HISTOGRAMS present the same information in diagrammatic form. The essential purpose with numerical data is to summarize a large mass of information in a compact and intelligible form.

The illustrations under FREQUENCY CURVE and FREQUENCY POLYGON are graphical examples of frequency distributions.

Frequency polygon.   A graphical depiction of a discontinuous (or discrete) FREQUENCY DISTRIBUTION, of which the frequency curve is a limiting form. Variate values are plotted against the abscissa and frequencies against the ordinate axis. When it is used for discrete value categories, the frequency polygon is constructed by plotting frequencies for each variate value and joining the points together in the manner shown in the diagram overleaf.

Where frequency polygons are used for interval measurements, it is common to use the *mid-point* of each interval measurement for plotting the frequency of that interval measurement. For example, in the above case the fact that twelve students achieved between 11 and 20 marks is shown as 12 against 15.5 on the abscissa. The twenty-one students who achieved between 21 and 30 marks are shown plotted as an ordinate value of 21 against 25.5 on the abscissa.

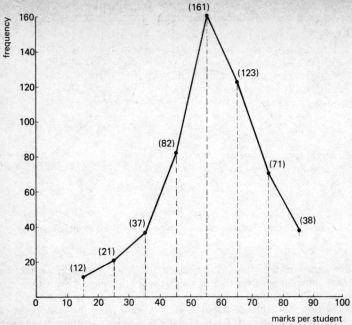

A frequency polygon showing the numbers of students achieving the marks 11–20, 21–30, 31–40, etc. in a school examination

Frequency table. A table which shows the distribution of frequency of an event or occurrence, categorized (where necessary) into sets of class interval values. For example, the discrete frequency distribution illustrated under FREQUENCY POLYGON may alternatively be shown as a frequency table.

Marks obtained (%)	Numbers of students
10 and under	0
11–20	12
21–30	21
31–40	37
41–50	82
51–60	161
61–70	123
71–80	71
81–90	38
91–100	0
Total	545

A frequency table showing the numbers of students achieving marks at specified intervals at a school examination

Friedman's test.   An adaptation of the CHI-SQUARED TEST to ranked
measurements to compare three or more matched samples. It was
devised by Friedman in 1937, and involves obtaining a chi-squared
statistic by squaring rank totals instead of summing the squares of
differences between observations and expected values. The sum of
squares of ranks is obtained (R) and the formula applied is:

$$\chi_r^2 = \frac{12R}{Nk(k+1)} - 3N(k+1)$$

where k is the number of observations, N is the number of values of
each observation, and χ_r^2 a symbol indicating the RANK chi-squared
coefficient.

Example
In three examinations, A, B and C, six candidates have the following
ranks. Is the difference between the examination results significant?

	candidates	1	2	3	4	5	6	
examinations	A	1	2	3	6	4	5	
	B	2	1	4	3	5	6	
	C	1	6	5	3	2	4	
total ranks (T)		4	9	12	12	11	15	
$R (= \Sigma T^2)$	=	16	+ 81	+144	+144	+121	+225	= 731

The table shows the rank totals for each candidate and the sum of
squares of those rank totals. The chi-squared statistic is therefore

$$\chi_r^2 = \frac{12 \times 731}{18 \times 7} - (9 \times 7)$$

$$= 6.62$$

Using tables which are available for non-parametric statistics, where
$k = 6$ and $N = 3$, the critical chi-squared values are 9.9 at 5 per cent
significance level and 11.7 at 1 per cent significance level. The test
statistic (6.62) is less than these values and therefore the difference in
examination performance is shown not to be significant at either
significance level. For the purposes of this test, ordinary chi-squared
tables may be used where k exceeds 7. Where, as in the above example,
k is either 7 or less, special tables are required.

Function.   The expression in mathematical terms or symbols of a relationship $(y = f(x))$ between a number of variables. For example, the
expression $y = a + bx$ is a mathematical function which gives the
relationship between y, a, b and x. A second example is a DISTRIBU-

TION FUNCTION where $F(x_1, x_2, \ldots x_p)$ is the cumulative frequency of observations at or below the variate measurements $x_1, x_2, \ldots x_p$.

Funnel sequence.   Even when the investigator is interested in a specific topic, it is often a good idea to begin the questioning on a much broader basis and gradually refine the questioning until it deals with the specific topics. This is called a funnel sequence of questions. For example, in market research, if the investigator's interest is in a particular brand of cereal, it may be advisable to ask some general questions on breakfast foods to begin with and lead into the questions about the brand. Otherwise, if the brand name were mentioned in the first question, answers to the other questions might be influenced by this and some otherwise useful information might be lost.

G

Galton ogive. An S-shaped form of CUMULATIVE FREQUENCY CURVE representing the normal distribution and resembling the OGIVE shape used in Gothic architecture. (◆ CUMULATIVE FREQUENCY POLYGON.)

Gamma distribution. A particular kind of distribution, also known as type III in the Pearsonian system of frequency distributions. It is continuously distributed from zero to infinity and is also unimodal. The formula for the frequency distribution is

$$dF(x) = \frac{e^{-x} x^{(\lambda - 1)}}{\Gamma(\lambda)} dx$$

Note the differences between the above formula and that of the POISSON DISTRIBUTION.

Gantt chart. A special kind of bar chart used in industrial and commercial statistics, in which periodic and cumulative periodic targets and attainments are shown.

Example
The following table gives a salesman's targets and attainments for a half year.

Month	Target	Actual	%	Cumulative target	Cumulative actual
January	3000	2500	83	3 000	2 500
February	2000	2250	112	5 000	4 750
March	2500	2500	100	7 500	7 250
April	2500	2750	110	10 000	10 000
May	3000	2250	75	13 000	12 250
June	3000	3000	100	16 000	15 250

In the chart below:
1. monthly and cumulative targets are shown for each month;
2. equal horizontal abscissa length does not necessarily mean the same

quantity, the important feature being whether the month's target has been achieved;

3. thin lines are used for monthly achievement values; and

4. a thick line is used for cumulative achievement.

January		February		March		April		May		June	
3000	3000	2000	5000	2500	7500	2500	10000	3000	13000	3000	16000

Monthly values

Cumulative values

15250

A Gantt chart measuring the performance of a salesman

Note that when a given monthly target is not achieved there is an apparent break in the upper (thin) line, and when it is occasionally exceeded, there is a small extension of the thin line underneath the main monthly line.

Gauss–Markov theorem. The basic theorem of least squares estimation. The theorem states that under certain conditions the Best Linear Unbiased Estimator (BLUE) of a population parameter is that derived by using LEAST SQUARES. 'Best' in this context means MINIMUM VARIANCE and 'linear' implies that the estimator is a linear combination of sample observations.

Geometric distribution. A discontinuous frequency distribution in which frequencies or relative frequencies are reduced in geometric progression as the values of a variate increase. The continuous form of this distribution is sometimes called an EXPONENTIAL DISTRIBUTION.

Example

Value (x)	Frequency (f)
1	128
2	64
3	32
4	16
5	8
6	4
7	2
8	1

The distribution should not be confused with that which appears normal when plotted on a horizontal logarithmic axis, of which the mode and median are the GEOMETRIC MEAN.

Geometric mean. A measure of average which consists of taking the nth root of the product of the n numbers being averaged. If x_1, x_2, ... x_n are the numbers to be averaged, then

$$\text{GM} = \sqrt[n]{(x_1 . x_2 . \ \dots \ . x_n)} \text{ or } (x_1 . x_2 . \ \dots \ . x_n)^{1/n}$$

It is often used to average percentages and ratios, and sometimes used in the construction of share price indices.

Geometric progression. A series (or sequence) of numbers ranked in rising or falling order such that the ratios between successive pairs of terms are constant.

Example
The series 1, 3, 9, 27, 81 ... is a geometric progression because the ratio between each successive pair (1, 3), (3, 9), (9, 27) is constant (= 3). The MEDIAN of the series is the GEOMETRIC MEAN (or geometric average). The sum of a geometric progression is

$$a \frac{r^n - 1}{r - 1}$$

where a is the first term of the series, r is the ratio between each successive pair of terms and n is the number of terms. Thus, in the above example, $a = 1, r = 3$ and $n = 5$, so that the sum of the terms is

$$1 \times \frac{3^5 - 1}{3 - 1} = \frac{242}{2}$$
$$= 121$$

Among other things, the concept of geometric progression is important in compound interest calculations for the computation of sinking-fund values, where the increment of a rate of interest (r) may be expressed as $(1 + i)$, i being the rate of interest, and therefore the sum of a sinking fund of n terms (i.e. $n - 1$ years) may be calculated using the formula

$$\frac{(1 + i)^n - 1}{i}$$

Gini coefficient. This gives a measure of inequality in the distribution of a variable throughout a population. If the variable is perfectly equally distributed, then each percentage of the variable is accounted for by the same percentage of the population and the Lorenz curve is a diagonal straight line. For progressive degrees of inequality, the Lorenz curve departs more and more from a straight line. The Gini coefficient of inequality is based on the area between the Lorenz curve and the diagonal straight line. In the diagram, the Gini coefficient is the shaded area expressed as a proportion of the triangular area below the diagonal.

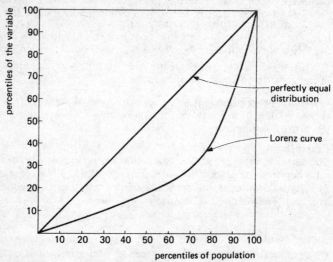

Gompertz curve (growth curve).   A trend curve representing either growth or decline where the rate of change either increases or declines at a constant percentage or rate. The equation of the curve is

$$Y = ae^{-bt}$$

where Y is the dependent variable, a is a constant and bt is the constantly increasing or declining rate of growth.

Goodness of fit.   A measure of the agreement between observed and expected values. The expected values may be determined by a regression model (in the case of REGRESSION ANALYSIS) or some other proposed or theoretical model. The extent of goodness of fit is measured by the extent to which expected values plotted on a given line or curve do not differ from observed values. Generally the lack of fit is measured in terms of some function of the sum of squared differences between observed statistics and expected values. The CHI-SQUARED TEST can be used to test goodness of fit in many cases.

Grade.   Galton used the word to mean the relative cumulative frequency of the variate value, or the proportion of total frequency having values less than or equal to the variate value. In the table under CUMULATIVE FREQUENCY, for example, 70 of the 250 office workers have salaries of £1999 or under. The grade of the value £1999 is therefore 7/25ths, or 0.28.

If the categories used in the cumulative frequency distribution are discrete, the calculation of grade is similar except that only half the frequency value of the highest variate class is included. Thus if total frequency is 1000 and variate ranges from 1 to 10, and if the cumulative frequency of variate values from 1 to 6 is 575 and the frequency of variate value 7 is 80, the calculation of grade for variate value 7 would be

$$\frac{575 + \frac{1}{2}(80)}{1000} = 0.615$$

Grand mean. A mean of a number of mean values. The term is used, *inter alia*, in analysis of variance and in mean range method of computing standard deviation. Thus, in the latter usage, a group of items is divided into subsets of a specified size (e.g. $n = 6$). Then the means and ranges of each of the sets are calculated. The mean of the ranges is determined and multiplied by a factor whose value depends on the number of items used in each subset. The result is an approximation of standard deviation. The parameter mean may be estimated by computing the mean of each of the means of the subsets. This mean of means is the grand mean.

Note that when the means are calculated from subsets of different sizes, they should be weighted to obtain the grand mean. Thus if the mean of a group of five observations is 8, and the mean of another group, consisting of fifteen observations is 12, the grand mean is

$$\frac{(5 \times 8) + (15 \times 12)}{20} = \frac{220}{20}$$
$$= 11$$

or, more simply

$$\frac{(1 \times 8) + (3 \times 12)}{4} = \frac{44}{4}$$
$$= 11$$

Grid. Two sets of parallel equally spaced lines drawn such that the sets of lines intersect each other vertically. A grid may consist of horizontal and vertical lines or of diagonally drawn intersecting lines. Grids may be used in area sampling to select particular subsets of an area for specific examination.

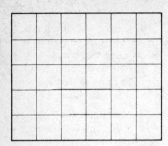 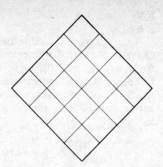

Examples of grids

Gross errors.   The total errors involved in the measurement or observation process. The examination of gross errors is particularly important in methodological studies where we are interested in the process of measurement itself. Whereas estimates of population means or totals are affected only by NET ERRORS – that part of the total error which does not cancel out over all the observations – MEASURES OF ASSOCIATION are affected by gross errors. In addition, any attempt to eliminate errors must be based on a knowledge of the gross errors involved and the circumstances under which they arise. (◆◆ RESPONSE ERRORS.)

Group.   A set of observations or of elements in a population which possess a common characteristic or set of characteristics.

H

Half-open interval. A procedure used to link elements not appearing in the population list with those that do in order to reduce NON-COVERAGE. We may, for example, suspect that all the dwellings in a street do not appear on our list. We therefore instruct the interviewer or enumerator when she arrives at a designated dwelling to check whether there are any unlisted dwellings between the designated dwelling and the next listed dwelling. If such dwellings exist, they should be included in the sample since they would otherwise have no chance of selection (i.e. they would have a zero probability of selection). In this way each unlisted dwelling is linked with a listed dwelling and is therefore allocated an appropriate probability of selection. The list itself must, of course, be unambiguously ordered if this method is to be applicable in practice. The term 'half-open interval' is derived from the fact that each designated unit represents an open-ended segment of the list – all units from the designated unit up to, but not including, the next listed unit.

Haphazard sampling. Selecting the sample for observation without taking particular care about the representativeness of the elements selected. Samples of volunteer subjects are haphazard samples. It is not appropriate to make inferences about the whole population on the basis of such samples. The term should not be confused with RANDOM SAMPLING, in which every element in the population has a known non-zero probability of selection.

Harmonic mean. A special kind of mean which is determined by calculating the mean of the reciprocals of the values concerned and then calculating the reciprocal of this quantity. The harmonic mean of 30, 40 and 50 is

$$\frac{1}{1/3\,(1/30\,+\,1/40\,+\,1/50)} \;=\; 38.3$$

A problem which may be solved by using the harmonic mean is given in the example.

Example

A car travels 10 miles at 30 miles per hour, a further 10 miles at 40 miles per hour and a subsequent 10 miles at 50 miles per hour. Its average speed for the 30-mile journey is not 1/3 (30 + 40 + 50) = 40 miles per hour, for the numbers of miles covered are equal but the times taken at different speeds are different. The correct computation of an average rate of miles per hour in these conditions would be the harmonic mean:

$$\frac{1}{3} \times \left(\frac{1}{30} + \frac{1}{40} + \frac{1}{50} \right) = \frac{47}{1800}$$

The reciprocal of 47/1800 is 1800/47 (i.e. 38.3 miles per hour).

Some care is needed in deciding whether the harmonic or arithmetic mean should be used. If the car had travelled for an hour at 30 miles per hour, and for two subsequent hours at 40 and 50 miles each respectively, the appropriate average would have been the arithmetic mean (40 m.p.h.), but if the values are expressed in distances the appropriate average is the harmonic mean (38.3 m.p.h.).

Heteroscedasticity. Scedasticity is a term, not much used, which denotes dispersion, particularly in terms of VARIANCE. The term heteroscedasticity is used in REGRESSION ANALYSIS to describe a situation where the variance of the error or disturbance term differs for different values of the independent variable. In general, the distribution of a variable is said to be heteroscedastic if its variance differs for different fixed values of another variable. If the variance is the same for all fixed values of the other variable, the distribution is said to be HOMOSCEDASTIC. For example, the plot below shows a case in which the disturbance term has a variance depending on the value of x, i.e. $y = a + \beta x + \varepsilon x$.

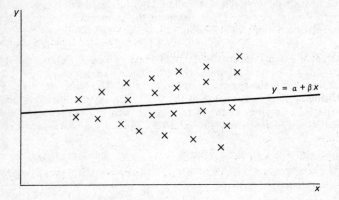

Hierarchy. The term hierarchy is generally used to mean a chart giving different levels of aggregation, for example:

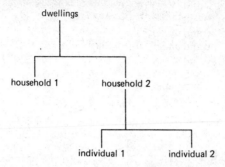

The term is also used to indicate a particular kind of correlation matrix where the highest correlation coefficients are found in the upper left-hand corner and the lowest correlation coefficients are found in the lower right-hand corner, and where there is a constant ratio between the elements of adjacent columns except for diagonal terms. The constituent coefficients obey the tetrad rule, for example:

$r_{13} \times r_{24} = r_{14} \times r_{23}$ in the matrix: $\begin{matrix} r_{11} & r_{12} & r_{13} & r_{14} \\ r_{21} & r_{22} & r_{23} & r_{24} \\ r_{31} & r_{32} & r_{33} & r_{34} \end{matrix}$

Assuming r_{13}, r_{14}, r_{23} and r_{24} to be:

$$\begin{matrix} 0.900 & 0.857 \\ 0.840 & 0.800 \end{matrix}$$

the cross products of the coefficient constituents are 0.72.

The existence of a hierarchy often indicates that a single common factor can account for the variation.

Histogram. The depiction of the frequency distribution of a discrete series of categories in the form of vertical bars. The area of each bar indicates the frequency (or relative frequency) of each category or class.

Example 1: Equal class intervals
The population of a block of terraced houses is divided into the following age classes:

Age in years	Frequency
0 and under 10	25
10 and under 20	27
20 and under 30	30
30 and under 40	15
40 and under 50	10
50 and under 60	6
60 and under 70	4
Total	117

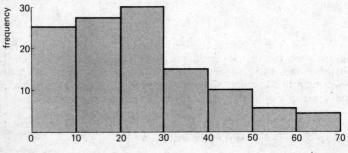

Example 2: Unequal class intervals

Vertical measurements should be adjusted when class intervals are not equal. For example, if it were known that there were fifty-seven people aged between ten and thirty but information on the age categories 10 and under 20 and 20 and under 30 was not available, the histogram would be drawn with a rectangle depicting '10 and under 30', but twice as wide as the other rectangles, with a height of 28.5 on the frequency scale. The area of this rectangle would represent the 57 people involved.

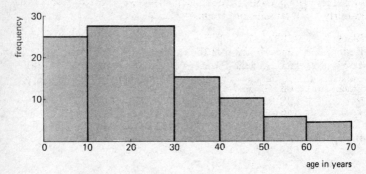

Historigram. A special form of HISTOGRAM where the abscissa is used for discrete time measurements or time interval measurements. The historigram illustrated below shows the numbers of books purchased by a particular county borough library system over the fiscal years 1971/2 to 1976/7.

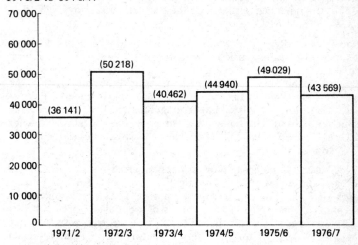

Homogeneous/homogeneity. The term is used in statistics in relation to samples from different populations which may or may not be identical. If the populations are identical, they are said to be homogeneous, and thus the samples drawn from these populations are also said to be homogeneous. It is more common to restrict the examination to particular characteristics of the populations, e.g. the means or VARIANCES. Thus a number of populations with identical means are homogeneous in their means.

For example, in a well-known investigation published in 1925, Fisher gives data on the numbers with black and red eyes respectively in thirty-three families of *Gammarus*.

Black	79	120	24	117	62	79	66	45	61	64	208	154	31	158
Red	14	31	6	29	17	20	12	11	14	13	52	45	4	45
Total	93	151	30	146	79	99	78	56	75	77	260	199	35	203

Black	21	105	28	58	81	25	95	47	67	30	70	139	179	129
Red	4	28	7	19	27	8	29	16	21	11	28	57	62	44
Total	25	133	35	77	108	33	124	63	88	41	98	196	241	173

Black	44	24	19	45	91	2565
Red	17	9	8	23	41	772
Total	61	33	27	68	132	3337

Using the CHI-SQUARED TEST, it can be shown that the ratios of black : red in the families do not differ significantly from the overall ratio 2565 : 772. In effect, 'All the families agree and confirm each other in indicating the black : red ratio observed in the total.' Thus the populations may be said to be homogeneous. If the test indicated that the ratios were not in agreement, then the populations would be said to be *heterogeneous*.

Homoscedasticity. Scedasticity is a term, not widely used, which denotes dispersion, particularly in terms of VARIANCE. The term homoscedasticity is used in REGRESSION ANALYSIS to describe a situation where the variance of the error term or disturbance term is the same for all values of the independent variable. In general, the distribution of a variable is said to be homoscedastic if its variance is the same for all fixed values of another variable. If the variance differs for different values of the other variables, the distribution is said to be HETEROSCEDASTIC.

Hypothesis. A supposition or assumption which acts as a foundation or as a starting point in an investigation, irrespective of its probable truth or falsity. In statistics the term hypothesis usually means an assumption concerning the parameters of a population from which a sample has been drawn. In the most usual case the hypothesis is a NULL HYPOTHESIS, e.g. that of no difference between the parent groups of two samples or between a sample and its assumed parent group. Methods of testing are employed to decide whether to reject or not to reject the null hypothesis. (◗ SIGNIFICANCE TESTS.)

I

Identical. Two expressions are identical if they are always equal to each other. The symbol $\equiv$ is used in equations where expressions on each side of the equation are identical to each other.

Incidence rate. The number of spells beginning during the observation period related to the average number of persons under observation. For example, if there are fifty men working in a factory and during one year there are seventy absences due to illness, the incidence rate of sickness absence is $70/50 = 1.4$ spells per man per year.

Incomplete coverage. A survey or census in which a considerable number of the elements in the population are not included may be called incomplete. The term incomplete CENSUS should be applied to a census which attempts to cover the whole population but fails to do so. Incomplete coverage should not be used to refer to situations where a properly designed sample from the population has been selected and observed. The difference between SAMPLE coverage and incomplete coverage is that in the case of incomplete coverage the exclusion of elements is of an arbitrary or accidental nature.

Independence (in probability). The probability of the outcome of event B is said to be independent of event A if the conditional probability of B given A is equal to the marginal probability of B, i.e. if $P(B|A) = P(B)$.

For example, if two dice are cast, A and B, the probability of obtaining a '6' from casting die B remains 1/6 irrespective of the outcome of casting die A. The probability of the one outcome is independent of that of the other. In contrast, if balls are drawn from an urn in succession, the probability of a particular ball being drawn at the second draw is dependent on the result of the first draw, given that the balls are not replaced. In this case, the two events, the two draws are not independent of each other.

If two events are independent of each other, the probability of a given pair of outcomes from the two events is the product of the marginal probabilities of each of the outcomes. Thus, for example, if an unbiased die is cast and an unbiased coin is tossed, the probability of obtaining a head and a '6' is the product of the probability of a head and that of a '6', i.e.

$$0.5 \times 0.167 = 0.0833$$

or

$$\frac{1}{2} \times \frac{1}{6} = \frac{1}{12}$$

Independence (of variates).   Two or more variates are independent of each other if pairs of readings from each of them are such that it is impossible to predict the value of one variate from the corresponding value of the other. In the bivariate normal distribution zero correlation between the two variates implies that they are independent of each other.

Independent variable.   A term used in REGRESSION ANALYSIS to mean one or a number of predictor or explanatory variables (or variates) from which a DEPENDENT VARIABLE is said to have resulted. It is a convention to use the letter Y to indicate the dependent variable and X_1, X_2, X_3, ... X_n for each of the independent variables. Thus a relationship between independent and dependent variables may be defined

$$Y = \beta_0 + \beta_1 X_1 + \beta_2 X_2 + \beta_3 X_3 \ldots + \beta_n X_n + \varepsilon$$

Because of the limited meaning of the word 'independent' in the context of the particular regression equation which is being considered, and because the word does not imply independence of other variates or of each other, some statisticians prefer to call independent variables explanatory variables, or REGRESSORS.

Index.   The term has a number of different meanings:
1. a subscript representing any value of a given variate, i.e. where X is a given variate and X_i may be used to represent any specific value, the subscript i is an index;
2. an exponent, i.e. a symbol denoting the number of times a particular quantity is to be taken as factor to produce the specified power;
3. a table of references, held in a computer memory in a particular sequence, such that it may be accessed to obtain items of data or filed information;
4.   a number used in computer processing for indexing, i.e. for retrieving information from an array of items in memory, whether on file or in a direct access store; or
5. in the context of INDEX NUMBERS, which compare by variations changes in the relative measurements of a variate over time or space.

Index numbers.   An index number is a quantity which, by its relation to a standard quantity determined at a base period, measures variation of a price or other magnitude over time or space.

Index numbers have been compiled in respect of magnitudes such as export prices, hourly rates of wages, import prices, industrial production, retail prices, retail sales, wholesale prices, volumes of exports and imports, and levels of trade. Cost of living index numbers are becoming increasingly important as a means of measuring inflation.

The essential features of an index number are: (1) its terms of reference (e.g. wages); (2) the unit of measurement, which may be prices or quantities; (3) the base period, with which other period measurements are compared; and (4) the weighting system used in the construction of complex indices, such as LASPEYRES and PAASCHE.

Example
The expenditure of a particular borough on books for its libraries during the period 1970 to 1975 is shown below. The year 1970/71 is used as the base period (3), the terms of reference are expenditure on libraries (1), the unit of measurement (2) on which the index is based is a monetary one and (4) the index is complex for it reflects both prices of books and quantities purchased, but the weighting system is implicit in the borough's choice of books for its libraries.

Year	Expenditure	Conversion to index number
1970/71	£67 075	100
1971/72	£71 671	$\frac{71671}{67075} \times 100 = 107$
1972/73	£76 754	$\frac{76754}{67075} \times 100 = 114$
1973/74	£83 348	$\frac{83348}{67075} \times 100 = 124$
1974/75	£92 857	$\frac{92857}{67075} \times 100 = 138$

Indirect effect.   That part of the effect of one variable on another in a PATH MODEL which operates through its effect on an intermediate variable in the model. For example, in the model below including father's occupation, son's education, and son's first occupation, part of the effect of father's occupation on son's first occupation may be brought about through the effect of father's occupation on son's education, which in turn affects son's first occupation. The rest of the TOTAL

EFFECT of father's occupation in this model is called the DIRECT
EFFECT.

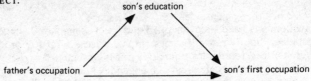

Individual response error.   The individual response error is the difference between the INDIVIDUAL TRUE VALUE and the information
recorded for the individual. (1) If, for example, the age of someone
born on 16 January 1946 is recorded on 16 January 1977 as thirty
years the individual response error is 1 year. The error could arise in a
number of ways. (a) The individual may have answered the question
correctly but the interviewer may have misheard or simply through
inattention written down 30. (b) On the other hand, the respondent
may have given false information deliberately or may have misunder-
stood the question. (2) If the interviewer asks how often the respondent went to the cinema in the previous month, the individual response
error may arise through a wish on the part of the respondent to impress
the interviewer or the respondent's memory may be at fault. Regardless of the source of the error the definition remains the same – the
difference between the information recorded and the individual true
value.

Individual true value (ITV).   Although usually defined in the context
of social surveys, this concept can be applied to any measurement
process. Its essential feature is that the value must be independent of
the way in which it is measured. For example, if we are interested in the
age of someone born on 16 January 1946 and we want to obtain the
information on 16 January 1977, there is a unique correct value – 31
years. No matter how we try to obtain the information the true value
remains the same. It is not always so easy to define this individual true
value. If we are measuring attitudes or opinions, for example, the
definition may be obscure. However, the concept is useful as an ideal at
which to aim and it is particularly important in the context of RESPONSE
ERRORS. Three criteria should be satisfied by the ITV: (1) it should be
uniquely defined; (2) it should be defined in such a way that the purposes of the survey are met; (3) it should be defined in terms of operations that can actually be carried out (even though these operations
may be very difficult or expensive to perform).

In practice, it may be very difficult to satisfy all three criteria simultaneously and some compromise may be necessary. It would be possible to define the ITV as the answer obtained to the question. This
would satisfy (1) and (3) above, but it is unlikely that we could accept it

as a satisfactory definition since it would almost certainly not satisfy (2).

Information.   A term which has a particular meaning in the theory of estimation.

The amount of information about a parameter φ from a randomly drawn sample of n independent observations, where the population from which the sample is drawn has a frequency function $f(x, \varphi)$ can be defined as

$$n \, \varepsilon \left(\frac{\delta \log f}{\delta \varphi} \right)^2 \equiv n \int_{-\infty}^{\infty} \left(\frac{\log f(x,\varphi)}{\delta \varphi} \right)^2 f(x,\varphi) \, dx$$

In general it may be said that if the variance of unbiased estimators of φ is large the information is small, and that if the variance is small the information is large.

An estimator φ is said to be efficient if equality exists where the Cramer Rao inequality is applied. If t is an estimator of φ with a bias $b(\varphi)$ and with a distribution frequency function $f(x, \varphi)$ then where $b(\varphi) = \varepsilon(t) - \varphi$ the inequality is that

$$\text{Var } t \geqslant \varepsilon \frac{(1 + \delta b/\delta \varphi)^2}{(\delta \log f/\delta \varphi)^2}$$

If there is equality the estimator is said to be efficient.

Input–output analysis.   A technique in which the output of each industry in an economy is analysed according to the use made of it. It can be used to calculate the level of output in different industries necessary to sustain a given pattern of final demand. Outputs are required both to satisfy final demand, and as intermediate goods used by industries in the production of other goods. An input–output matrix A is constructed whose (i, j)th element is the amount of the product of industry i (input) needed to produce a unit of product (output) of industry j. This matrix embodies the technological constraints on the economy. Suppose the vector x measures the outputs of the various industries, then Ax of this will be required as intermediate goods leaving the difference y as final demand, i.e. $(I - A)x = y$ or $x = (I - A)^{-1} y$. Thus given y and A one can calculate the output levels in the various industries.

Interaction.   When the effect of a change in one variable differs for different levels of another variable there is said to be an interaction between them. A simple example may illustrate the concept. The level of absenteeism of a worker is related both to the worker's sex and the worker's family responsibilities. In a particular study men may be absent more often than women, However, the effect of family responsibilities may be different for men and women. For men, increased family responsibilities may decrease absenteeism. For women, however, the opposite is the case: the greater the family responsibilities the greater the absenteeism.Thus sex and family responsibilities interact with one another in terms of the variable absenteeism.

Interdecile range.   The range of observations of a variate between the first and ninth deciles. Its advantage over QUARTILE RANGE is that it takes account of some extreme values and is a valid measure of dispersion, but omits those values which are so exceptional as to be lower than the first decile or higher than the ninth decile.

Example
In a set of forty-nine observations of a variate, the extreme ten observations are:

Rank	Observation	Rank	Observation
1	2	45	109
2	41	46	110
3	90	47	150
4	95	48	600
5	98	49	900

The absolute range $(900 - 2 = 898)$ contains some extreme values, while the interquartile range would, in this case, be small. The interdecile range $(109 - 98 = 11)$ is a compromise between the two measures and has some of the advantages of both.

Interpenetrating samples (subsamples).   With interpenetrating sampling, instead of selecting a single sample from the population we select a number of SUBSAMPLES using the same sample design for each. Each subsample must itself be a self-contained sample of the population. Interpenetration may be used with any sample design and at any stage of the selection. In general, there are two distinct purposes for which interpenetration is used. First, different operations or methods may be used in the data collection or processing stages in each of the subsamples. For example, a different interviewer may be allocated to each of the subsamples in an interviewer survey; this enables us to compute estimates of INTERVIEWER VARIANCE. Secondly, the estimates obtained from the subsamples may be used to provide a simple method of calculating SAMPLING ERROR. (◆◆ REPLICATED SAMPLING.)

Interquartile range.   A common measure of the dispersion of values of a set of numbers. It is the difference between the upper quartile and the lower quartile. For example, in the set 1, 2, 3, 5, 7, 8, 9, 14, 15 the upper quartile is 9, the lower quartile is 3 and the interquartile range is $9 - 3 = 6$. (◆ QUARTILE RANGE.)

Interval estimate. An ESTIMATE, based on sample observations, of a population parameter or characteristic which is given in the form of a range of values rather than a single value (as in POINT ESTIMATION). For example, in estimating the proportion of electors who will vote Labour in the next election we might express the estimate as 40% ± 4%, i.e. the range 36% − 44%. We would usually associate a probability or confidence level with the interval. Thus we might say that we are *95% confident* that the observed interval will contain the population mean value. This is saying in effect that for 95% of possible samples our interval estimate will cover or include the true value.

Interval scale. A SCALE OF MEASUREMENT on which the scores are based on equal units of measurement. It is possible to compare scale scores on an interval scale not merely in terms of the order of the scores but also in terms of the distances between them. The Fahrenheit temperature scale is an example. Thus the difference between 65°F and 55°F is 10°F, which is the equal to the difference between 42°F and 32°F. But 60°F is not twice as hot as 30°F. For an interval scale, the numerical scores may be changed by adding or subtracting a constant or by multiplying each score by a constant without changing the meaning or usefulness of the scale. We could, for example, subtract 32 from each Fahrenheit temperature scale score and multiply by 5/9. This would, in fact, give us the Celsius temperature scale which has exactly the same properties and provides an equivalent amount of information.

Almost all statistical techniques can be used in analysing data measured on an interval scale.

Interview. A structured social interaction between two people – an interviewer whose task it is to obtain information, and a respondent whose function it is to provide this information. The two principal characteristics of an interview are, first, that it is governed by the ordinary rules of social situations and, second, that it is task-orientated, the task being the giving and obtaining of information. The degree of structure will depend on the purpose of the interview.

Interviewer bias. Bias in the responses or recorded information which is caused directly by the interviewer. It may be due to systematic errors in recording the answers; failure to contact the designated individuals; or to a systematic influence on the way in which the respondents answer the questions. The latter situation can arise in a number of ways. First, the age, sex, education or social class of the interviewer may cause the respondent to modify his answers to some questions. Second, the interviewer may form an impression of the respondent early in his interview and may allow this impression to influence either the way in which later questions are asked or the way in which ambiguous answers are coded. Third, the opinions and beliefs of the interviewer may communicate themselves to the respondent and influence the way in which the questions are answered.

Each interviewer has an average 'interviewer bias' on the responses in his workload. Different interviewers may introduce different biases into the responses. The average bias over all the interviewers in the survey is the net interviewer bias. The rest of the effect is included in the INTERVIEWER VARIANCE.

Interviewer effect.   A term sometimes used to denote the totality of errors in survey responses due to the interviewer. It includes therefore INTERVIEWER BIAS and INTERVIEWER VARIANCE.

Interviewer–respondent interaction.   The two participants in an interview are the interviewer and the respondent. It is important that proper relations (*rapport*) should be established by the interviewer at the beginning of the interview. Otherwise imperfect or inaccurate information may be given by the respondent. By establishing appropriate conditions for the interview at the outset the interviewer may reduce the extent to which interaction between his characteristics and those of the respondent lead to errors in the responses.

Interviewer variance.   Each interviewer has an individual INTERVIEWER BIAS on the responses in his workload. Even if these biases cancel out over all the interviewers in the survey, their presence has an effect on the precision of the estimate obtained. Thus the effect of the variation between the individual interviewer biases is called the interviewer variance. The effect can be considerable particularly for attitudinal questions. It can be summarized by the INTRACLASS CORRELATION COEFFICIENT. This is a measure of the similarity between the responses obtained by an interviewer within his workload which is caused by the way in which the interviewer influences the respondents. It may also be described as the proportion of the total variance which is accounted for by the differences between interviewers.

Intraclass.   Within a class or group. We distinguish between intraclass and interclass. In the analysis of variance, for example, the total sum of squares can be partitioned into the interclass (as between groups) sum of squares and the intraclass (or within group) sum of squares. (◆◆ INTRACLASS CORRELATION COEFFICIENT.)

Intraclass correlation coefficient.   A measure of the internal homogeneity of a class, category, group or family of elements. The intraclass correlation coefficient indicates how similar the elements are within classes compared to the overall similarity of the elements within the population as a whole. In a one-way analysis of variance, the coefficient indicates the proportion of the total variance attributable to the variability between classes. The larger the coefficient, the more homogeneous the elements within classes – the extreme value of $+1$ for the coefficient indicates that all the elements in each class have the same value, i.e. all the variance is accounted for by the variation between classes. In other words, the sum of squares within classes is zero.

Intracluster correlation coefficient. A synonym for INTRACLASS CORRELATION COEFFICIENT. The term 'intracluster' is used mainly in connection with survey sampling where the 'clusters' are the groups of elements selected in CLUSTER SAMPLING. The coefficient is usually positive for human populations, since people living in the same neighbourhood tend to be more similar to one another than they are to people living in other neighbourhoods. A high positive intracluster correlation coefficient implies that within-cluster variability is relatively low – the clusters are relatively homogeneous.

Inverse. The inverse of square $p \times p$ matrix A is written A^{-1}, and is a $p \times p$ matrix such that

$$AA^{-1} = A^{-1}A = I_p$$

where I_p is the unit matrix of order p which consists of ones on the main diagonal and zeroes elsewhere; for example, the unit matrix of order 5 is:

$$I_5 = \begin{bmatrix} 1 & 0 & 0 & 0 & 0 \\ 0 & 1 & 0 & 0 & 0 \\ 0 & 0 & 1 & 0 & 0 \\ 0 & 0 & 0 & 1 & 0 \\ 0 & 0 & 0 & 0 & 1 \end{bmatrix}$$

Inversion. If we define a standard order of elements (e.g. the natural order of numbers 1, 2, 3, 4, 5, ...), an inversion of two elements means that they are in the opposite order to that given in standard order. Thus, compared to 1, 2, 3, the order 1, 3, 2 contains one inversion (3, 2 instead of 2, 3). The concept is used in the calculation of KENDALL'S TAU.

Inverted funnel sequence. In asking questions about broad issues, it is difficult to ensure that the respondent has considered the components of these issues. One method of dealing with this problem is to ask questions first about some of the specific issues in order to lead all the respondents to base their answer to the broad question on a consideration of the major issues involved. This is called an inverted funnel sequence of questions. For example, if we are interested in people's judgement of the Government's overall performance, questions on particular economic and social issues might precede the general question.

Isometric chart.   An attempt to illustrate three-dimensional material, illustrating three variates, on a two-dimensional graph. Conventions are employed to make the two-dimensional graph appear three-dimensional. Axes follow a box pattern, one axis being vertical, and two non-horizontal axes representing the 'floor' of the box.

Example
A common use of isometric charts is in the economics of production, where the two dimensions comprising the 'floor' of the box may be used to illustrate two inputs, such as manhours and materials, and the vertical axis is used to illustrate the volume of output of one product. The following isometric chart shows levels of production of a printer, whose two major inputs are labour and paper, and whose one major output is measured in quantities of books of a standard size. The shading illustrates the position where L units of labour and P units of paper are used to manufacture B units of books.

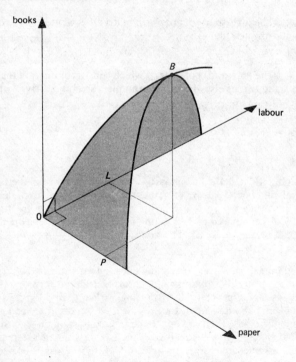

J

Joint regression. A form of regression where one of the regressors is a cross-product, or other function, of the other regressors or of some of them.

Example

$$y = \beta_0 + \beta_1 x_1 + \beta_2 x_2 + \beta_3 x_3 + \varepsilon$$

where x_3 is a cross-product (i.e. $x_1 x_2$).

Judgement sampling/purposive sampling. Any method of sampling where instead of using PROBABILITY SAMPLING the investigator uses his own 'expert' judgement to select elements which he considers to be typical or representative of the population. The problem with such methods is that there is no guarantee that the expert will not introduce serious biases into the results by basing the selection on faulty premises. In general, probability sampling is to be preferred.

K

Kendall's tau. A rank correlation coefficient based on two sets of rankings of a collection of objects or individuals. It is measured by examining how many inversions of order of pairs or objects of one set of rankings is required to make it agree with the other.

For example the two orderings A B C and A C B can be made to agree with one another by inverting the order of C and B in the second set, i.e. one inversion. The value of the coefficient is given by

$$\tau = 1 - \frac{2s}{\frac{1}{2}n(n-1)}$$

where s is the necessary number of inversions and n is the number of objects or individuals. The value of the coefficient in this case is therefore:

$$\tau = 1 - \frac{2 \times 1}{\frac{1}{2} \times 3 \times (3-1)}$$

$$= \frac{1}{3} \text{ or } + 0.33$$

Kruskal and Wallis test. An extension of the Wilcoxon Sum of Ranks test to cases where there are more than two sets of measurements, thus comparing three or more unmatched random samples of measurements.

Example
On a particular radio station, four broadcasts are given on behalf of the Conservative party, five on behalf of the Labour party and three on behalf of the Liberal party. The table shows their duration times in minutes.

Party	Duration time (*mins*)				
Conservative	30	20	18	14	
Labour	22	34	31	21	17
Liberal	40	15	19		

Procedure
1. Prepare a table ranking and analysing the ranks of all observations.

Observation	Rank value	Conservative	Labour	Liberal
14	1	1		
15	2			2
17	3		3	
18	4	4		
19	5			5
20	6	6		
21	7		7	
22	8		8	
30	9	9		
31	10		10	
34	11		11	
40	12			12
	$n = 12$	$r_1 = 20$	$r_2 = 39$	$r_3 = 19$
		$n_1 = 4$	$n_2 = 5$	$n_3 = 3$

2. Apply the formula

$$\chi^2 = \frac{12}{n(n + 1)} \left(\frac{r_1^2}{n_1} + \frac{r_2^2}{n_2} + \frac{r_3^2}{n_3} + \ldots \right) - 3(n + 1)$$

where n = total number of ranks, n_i the number of ranks for each category and r_i the sum of ranks for each category.

3. In this case

$$\chi^2 = \frac{1}{13} (100 + 304.2 + 120.3) - 39$$

$$= 1.35$$

4. The lowest value of χ^2 where unmatched categories consist of 3, 4 and 5 observations (even when $P = 10\%$) is 4.5, and as 1.35 does not exceed 4.5 there is no evidence of significant difference in the time allocated to each of the political parties.

Note. A table of χ^2 values for the Kruskal and Wallis test appears in the *Journal of the American Statistical Association* 1952, pp. 614–17. Reproduction is not possible in an entry of this size because of the large number of combinations of unmatched categories n_i.

Kurtosis.   The extent to which a unimodal frequency curve rises sharply at the mode. If it is sharply peaked at the centre, the curve is leptokurtic. If the central section of the curve is flattened or rounded it is platykurtic. An intermediate degree of peakedness such as that of the NORMAL DISTRIBUTION is called mesokurtic. The accompanying diagram illustrates the three curves.

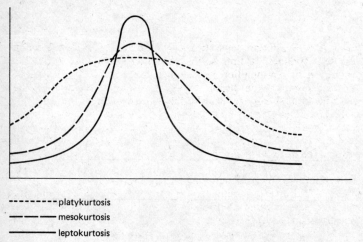

------- platykurtosis
— — — mesokurtosis
———— leptokurtosis

L

Lag. A period of time separating a causal set of values from a dependent set of values. In time series analysis, for example, a series of production or consumption figures may be associated with national budget decisions taken during the previous year. The variate values which illustrate the budget decisions may be regarded as indicating the causal factor, and those which illustrate production and consumption a year later are lagged dependent variables. Similarly demand for a product may be affected by a price change, but there is usually a lag between the decision to change price and the rise or fall in the quantity of the product demanded.

Sometimes there are a series of lags in the effect of an action. At time t, for example, a reduction of prices in one store will effect increased sales during a subsequent week ($t + 1$), but if the price reductions are advertised over a wide area the effect may be distributed over a number of weeks ($t + 1, t + 2, t + 3, \ldots t + n$). Note that the term is sometimes used more generally to indicate the interval lapse between, for example, the data relating to a particular month and its publication. (♦ DISTRIBUTED LAG.)

Laspeyres index. Probably the most popular form of index number, the Laspeyres index uses weights of a base period to lead the components. For example, in a price index the quantities of goods purchased in the base period are valued in terms of prices prevailing in other periods and compared with their value in the base period. Let $q_i^{(0)}$ be the base period quantities purchased at prices $p_i^{(0)}$, and let $p_i^{(1)}$ be the prices prevailing in some other period of interest ($i = 1, 2, \ldots n$). Then the Laspeyres index is given by

$$\frac{\sum\limits_{i=1}^{n} p_i^{(1)} q_i^{(0)}}{\sum\limits_{i=1}^{n} p_i^{(0)} q_i^{(0)}} \times 100$$

For example, suppose that in 1974 consumers buy on average 4 pints of beer at 15p a pint, 2 tots of whisky at 18p each, 1 glass of gin at 17p and 3 glasses of wine at 12p each. Suppose, further, that in 1978 these prices had changed respectively to 20p, 26p, 25p and 15p. Then the Laspeyres alcohol price index for 1978 based on 1974 = 100 is

$$\frac{4 \times 20 + 2 \times 26 + 1 \times 25 + 3 \times 15}{4 \times 15 + 2 \times 18 + 1 \times 17 + 3 \times 12} \times 100 = \frac{20200}{149}$$

$$= 135.6$$

showing an average price increase of 35.6%.

An alternative index – weighted by current period quantities – is the PAASCHE INDEX.

Latin square, Latin square design. A term used in the design of experiments for a design which attempts to remove the experimental error resulting from the variation from two sources which can be identified with the rows and columns of the square by permutation of the treatments, each of which is identified with a Latin letter.

For example, let us assume that there are four treatments denoted by the letters A, B, C and D and that there are two causes of experimental variation (e.g. time and place). In order to minimize the number of experiments it is necessary that, if the two causes of experimental variation are denoted by the rows and columns of the square, each letter (representing a treatment) appears once and only once in each row and column of the square. In this case a specimen Latin square which could be used for experimental design is

A	B	C	D
B	A	D	C
C	D	A	B
D	C	B	A

Law of large numbers. The law is a popular interpretation of the principle of statistical regularity, and states that large groups or aggregates of data show a greater degree of stability than small ones. Thus if a reasonably large sample of items is drawn at random from a parent group (or population) it will probably show the characteristics of the parent group more representatively than a small sample would do. In fact, the sampling error is inversely proportionate to the square root of the number of items in the sample.

The theoretical justification of the Law of Large Numbers is to be found in the CENTRAL LIMIT THEOREM.

Least squares estimate. An estimate obtained by a least squares process or method. For example, in least squares REGRESSION ANALYSIS a line of best fit is obtained and the relationship between observations of a dependent variable and the regressors is expressed as an equation. The exact points on the line corresponding to the observations are estimates, not in the usual statistical sense (e.g. an estimate of a parameter from a statistic) but in the more general sense, that they are values produced by a rule of estimation (in this case the least squares regression model or equation). For example, in the simple diagram below, if ABCDE is the line of best fit and X, Y, and Z are observations, then B, C and D are least squares estimates. The line is such that the sum of squares $(X - B)^2 + (Y - C)^2 + \ldots$ is least. In a simple regression model using analysis of variance, the sum of squares of errors $(X - B)^2 + (Y - C)^2 + (Z - D)^2$ is used to estimate the stochastic element in the model. ($\blacklozenge$ ERROR SUM OF SQUARES.)

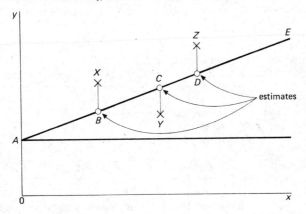

Least squares estimator. In statistical analysis an ESTIMATOR is a rule by means of which the parameters of a parent group may be estimated, the values produced being termed ESTIMATES. In any technique using LEAST SQUARES METHOD (as, for example, least squares regression analysis), the estimator is the rule or model obtained by least squares (for example, that y should be estimated as a + bx); and the estimate is the value produced by the rule.

Least squares method. A method of estimating a function or line of best fit by employing the principle that the optimal function is that which minimizes the squares of differences between the ESTIMATES which it produces and the OBSERVATIONS on which it is based. For example, in the diagram under LEAST SQUARES ESTIMATE the line ABCDE is the line of best fit if, and only if, it minimizes the ERROR SUM OF SQUARES $(B - X)^2 + (C - Y)^2 + (D - Z)^2$.

The least squares method under certain conditions provides the MINIMUM VARIANCE ESTIMATOR in the class of linear unbiased estimators. If the stochastic (or disturbance) terms in the regression model (ε) are normally distributed, least squares estimation is equivalent to MAXIMUM LIKELIHOOD estimation.

Level of measurement.   A term used to denote the strength of the SCALE OF MEASUREMENT on which the data are measured. The lowest level of measurement is the NOMINAL SCALE for which the only applicable statistical methods are those based on frequencies or counting – the mode, contingency tables, etc. Next comes the ORDINAL SCALE to which the whole class of NON-PARAMETRIC METHODS can be applied. INTERVAL SCALES and RATIO SCALES permit the use of all statistical techniques. In the physical sciences much of the measurement is at the interval and ratio levels. In the social sciences, particularly when measuring attitudes, we must be satisfied with measurement at the ordinal level.

Likelihood.   A measure of the plausibility of a postulated population structure given values of observations on a random sample from that population. The postulated population structure is usually in terms of a model for the distribution function of a random variable defined on that population together with values for parameters of that function. The probability of observing the particular set of sample values of the random variable for the postulated structure is its likelihood. Likelihood can be evaluated for alternative population structures and the plausibility of each compared.

For example, a population could consist of the hypothetically infinite tosses with a particular coin. It may be postulated that the coin is a fair one and that the proportion of heads, p, in this population would equal $\frac{1}{2}$. A random sample of ten tosses is made and it is supposed that the outcome is six heads. From the BINOMIAL DISTRIBUTION the likelihood is

$$\frac{10 \times 9 \times 8 \times 7}{4 \times 3 \times 2 \times 1} \ (\tfrac{1}{2})^6 \ (\tfrac{1}{2})^4 = 0.205$$

Likelihood function.   This is likelihood expressed as an algebraic function of possible values of parameters for a particular population distribution and the observed set of sample data.

For example, in considering the population of outcomes of tossing an identical coin under identical conditions a hypothetically infinite number of times, the parameter p is the long run proportion of heads or the probability that a head will appear on each toss. In a sample of ten suppose six heads appeared. From the binomial probability that six heads will appear, the likelihood function for p will be

$$L(p) = \frac{10 \times 9 \times 8 \times 7}{4 \times 3 \times 2 \times 1} \ p^6 \ (1 - p)^4$$

The likelihood function, though derived from the probability that the observed sample data will appear for specific values of the parameter, is not itself a probability function. It is a function of the possible values of the parameter(s) given the sample data.

Likelihood ratio. The ratio of the likelihood of one postulated population structure to another for a given set of sample data. Likelihood ratios form the basis for likelihood ratio tests.

Likelihood ratio tests. These are tests of null hypotheses of a population structure against specified alternative hypotheses based on the likelihood ratio. Usually the null hypothesis is simple in that it specifies the population distribution and values of parameters. The likelihood of the null hypothesis given the sample data can be found by substituting these values in the likelihood function. The alternative hypothesis may sometimes be simple in that it specifies particular values for the parameters. More usually, however, it will be compound in that it specifies a subset of possible values of the parameters. In the former case the likelihood of the alternative can be found and hence the likelihood ratio of the alternative hypothesis to the null. In the latter case the most plausible parameter values in the subset specifying the alternative are found by maximizing the likelihood function within this subset. The likelihood ratio is formed using these parameter values. The resulting number is called the likelihood ratio test statistic.

For example, a null hypothesis about the population of tosses of a coin might be that the coin is fair, i.e. that $p = \frac{1}{2}$. A simple alternative could be $p = 0.7$, a compound alternative that the coin was biased towards heads and $p > \frac{1}{2}$. Under the simple alternative the value of the likelihood ratio test statistic would be if six heads were observed in a sample of ten tosses: $(0.7)^6 (0.3)^4/(0.5)^6 (0.5)^4$.

Under the compound alternative the value of p which maximizes the likelihood function, in this case $p = 0.6$ would be substituted in the likelihood ratio and the value of the test statistic would be: $(0.6)^6 (0.4)^4/(0.5)^6 (0.5)^4$.

The larger is the value of the likelihood ratio test statistic the more plausible is the alternative hypothesis compared to the null hypothesis. Consequently it forms a suitable basis for a test procedure. Under the usual rules for constructing statistical hypothesis testing procedures it would be required to find the probability sampling distribution of the likelihood ratio statistic under the null hypothesis, and this hypothesis would be rejected when its value exceeded a critical value at the specified significance level. In other words, when the relative plausibility of the alternative exceeded a critical value the null hypothesis would be rejected.

Though it is not immediately obvious, many of the more common testing procedures in statistical methods are likelihood ratio tests or are based on simple functions or variants of the likelihood ratio test statistic.

Linear, linearity. A locus of points is said to be linear or display linearity if it takes the form of a straight line. The term is more often used in elementary statistics in the context of LINEAR REGRESSION models and LINEAR PROGRAMMING. Linear regression models assume that the relationship between two variates is of the form $Y = a + bX$ where a and b are constant. A simple case is illustrated in the graph below.

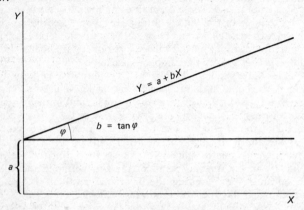

Similarly linear programming models are those where the constraints and objective function are depicted in graphical formulation by means of straight lines.

Linear estimator.   An estimator which is a linear function of the observations. The sample mean is a linear estimator of the population mean, for example, since it can be written in the form

$$\bar{x} = \frac{1}{n} \sum_{i=1}^{n} x_i$$

where the x_i ($i = 1, \ldots n$) are the sample observations.

Linear program (linear programming).   A branch of mathematical programming which is concerned with making quantitative decisions when several sets of differing resources and alternative courses of action are involved in the decision process. In the case of linear programming all the functions defining objectives (ends) and constraints (restrictions or limitations of means) are linear. Linear programs use iterative, stepwise procedures for determining the optimum solution in any particular case.

Example
Two restricted inputs, paper and labour are used to produce newsprint and books. Each batch of newsprint sold adds £4 to the firm's profits and each batch of books (100) sold adds £5 to its profits. In a given period paper obtainable is restricted to 200 units of which 5 are needed

for each batch of book production and 3 for each batch of newsprint, while 600 hours of labour are available of which 7 are required for a batch of newsprint and 4 for a batch of books. As profit is to be maximized the programme may be written:

Objective: Maximize $4x_1 + 5x_2$ (profit from each form of output)

Constraints: Subject to $3x_1 + 5x_2 \leqslant 200$ (paper constraint)
$7x_1 + 4x_2 \leqslant 600$ (labour constraint)
$x_1, x_2 \geqslant 0$ (solution must not be negative)

The problem may be solved by graphical means or by the SIMPLEX METHOD. (◆◆ DUAL THEOREM (DUAL PROBLEM).)

Linear regression. A regression relationship between one or more independent variates (X_j) and a dependent variable (Y) where there is an expectation that the dependent variable is a first-degree function of the independent variate(s) (or regressor(s)).

For example, the expression $Y = a + bX$ is a case of linear regression. One simple and exact illustration of linear regression is that of an electricity account where there is a constant (a) payment for a meter usage and a cost (b) for each unit of electricity. If the meter charge is £3 per quarter and each unit of electricity is sold at 2p there will be an exact linear relationship between cost (Y) and usage (X), i.e. $Y = 3.00 + 0.02X$, such that if usage is 1 unit, cost is £3.02, while if usage is 10 000 units the cost is £203.00.

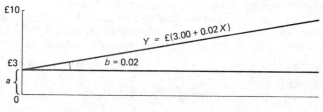

This is a case of perfect correlation between usage and cost. In most linear equations there is provision for an unexplained residual term ε (i.e. $Y = a + bX + \varepsilon$) which has zero expectation, but which may arise either from a random factor or for some other reason.

Linked samples. Two samples of the same size in which there is a one-to-one correspondence between their respective SAMPLE UNITS. The linking method may provide either an absolute or partial constraint on the selection insofar as the selection of a unit in one sample may determine uniquely the element to be selected in the other, or may simply restrict the choice of the corresponding element.

Logarithm. The index or power to which a number (on which the logarithm is based) may be raised, indicating the number of times which the base number is multiplied by itself. Logarithms are usually tabulated as means of abridging calculation, for multiplication and division may be achieved by addition and subtraction of logarithms, and powers and roots of numbers may be obtained by multiplication and division of logarithms.

Originally, logarithms were natural, or Napierian, based on e ($\approx$ 2.718). Hence the natural logarithm of 2.718 is 1.000 (i.e. 2.718^1), that of 5 is 1.609 (i.e. $2.718^{1.609}$), that of 7.382 is 2.000 (i.e. 2.718^2), etc. The quantity e was chosen as a base by Napier, whose discovery led to the popular usage of logarithms, because it proved to be the most accurate base from which logarithms could be calculated. Common usage, however, dictated a simpler base, and tables were later produced indicating values of logarithms to the base of 10. Hence the logarithm of 10 is 1.000 (10^1), that of 100 is 2.000 (10^2), that of 2 is 0.3010 ($10^{0.3010}$), and numbers are multiplied by aggregating their logarithms (e.g. $2 \times 2 = 10^{0.3010} \times 10^{0.3010} = 10^{0.3010 + 0.3010} = 10^{0.6020} = 4$) and division is effected by subtracting logarithms (e.g. $10/2 = 10^1/10^{0.3010} = 10^{1-0.3010} = 10^{0.6990} = 5$).

Before the advent of pocket calculators and computers logarithms formed the easiest available method of rapid multiplication and division, and the logarithmic scale forms the basis of slide-rule measurement.

Logarithmic scale. A chart drawn to logarithmic scale in one or both of its axes uses the intervals and gradations of equivalent logarithmic distances instead of normal ordinate and co-ordinate measurements. The most common graphical usage of logarithmic scale is the semi-logarithmic (or ratio-scale) graph, which uses logarithmic distances for

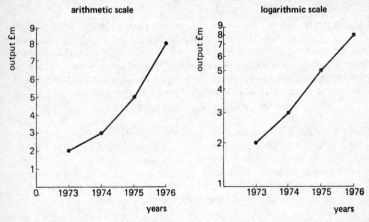

its measurements on the ordinate scale and normal distances on the abscissa. This kind of graph is useful for measuring rates of change. If one wishes to compare the output of one company with the comparative output of a group of companies relative movements may be easily indicated and equal vertical distances will represent equal proportionate changes. The two illustrations show the same information represented on arithmetic and logarithmic scales.

The arithmetic scale graph shows that there is the greatest absolute increase in output between 1975 and 1976, but the logarithmic scale graph shows the largest proportionate increase to occur between 1974 and 1975. Note that the comparison between variables is illustrated by the direction of the lines. Zero and negative measurements are not indicated, and a base line (as such) is therefore not necessary, although abscissa measurements are, of course, indicated on the lowest line of the graph.

Longitudinal surveys, longitudinal studies, panel surveys.   A longitudinal survey or panel survey is a survey in which information is sought from the same samples of respondents on more than one occasion. The study begins as a sample survey of the survey population. The selected individuals are then approached at intervals, by mail, by telephone or by personal interview. Longitudinal surveys provide a good opportunity for studying trends and also the effects of introducing particular measures (such as legislation or advertising). In addition, they make it possible to observe not only overall changes over time, but also which individuals change, and why they change.

The main problems with longitudinal studies are: (1) the loss of members of the sample over time; (2) 'panel conditioning' – members of the sample may be affected by the fact that they are continually being questioned about a particular topic or area; even (3) interviewer conditioning may become a problem.

Lorenz curve.   A graph designed to show the extent of inequality in the distribution of a variable throughout a population. Percentiles of the population are plotted against percentiles of the variable concerned. For example, if the lowest x% of the population earn only y% of total income (x, y) is a point on the curve. The Gini coefficient is a measure of inequality based on this curve. ($\blacklozenge$ GINI COEFFICIENT.)

M

Mann–Whitney U-test. This is a test of significance which is applicable to two independent samples where measurement is on at least the ordinal level. The test statistic U assesses the probability of a particular pattern of results in relation to the total number of possible patterns. The test assumes that the two samples are independent and that sampling is random. The HYPOTHESIS being tested is that there is no difference between the populations from which the samples were drawn.

Example

If two samples of children in primary schools are selected – one sample of children who have been to nursery school, the other of children who have not – we might want to ascertain whether, on the basis of an IQ test, children with nursery school education score more highly than children who have not been to nursery school. The hypothesis to be tested is that nursery school education makes no difference.

We calculate the test statistic U (for procedure see, for instance, Siegel (1956)) and look it up in the statistical tables. This gives a value of p.

Result: The probability of obtaining a value of U as small as the one observed if the hypothesis is correct is equal to p. Hence the larger the value of p the less likely that the hypothesis is false. If, however, $p < 0.05$ we reject the hypothesis at the 5 per cent level (i.e. we believe that the evidence shows that nursery school education does have an effect on IQ). If $p < 0.01$ we reject the hypothesis at the 1 per cent level. Otherwise we say that there is insufficient evidence to reject the hypothesis.

Marginal distribution. The distribution of row totals or column totals in two-way or in multiway tables.

For example in MARGINALS the distributions indicated in the marginal row and column

2, 10, 26, 41, 26, 12, 2; and
1, 9, 25, 40, 31, 10, 3

are marginal distributions of frequencies.

Similarly, in the entry MARGINAL PROBABILITY, the probability distributions indicated in the marginal row and column

0.60, 0.40; and
0.70, 0.30

are marginal probability distributions.

Marginal probability. In a two-way table, showing the probabilities of outcomes of two events, the row and column totals provide the separate probabilities of outcomes of each of the events.

Example
The following table shows the probabilities of selecting British and non-British males and females from a group of people in a particular town.

| Sex | Nationality | | Marginal probability |
	British	Non-British	
Male	0.56	0.14	0.70
Female	0.04	0.26	0.30
Marginal probability	0.60	0.40	1.00

The marginal probability of a female is thus 0.30, irrespective of whether the female is British or non-British. On the other hand, marginal probability may be used to calculate conditional probability. For example, the conditional probability that a person is female given that she is British is

$$\frac{\text{The probability of a British female}}{\text{Marginal probability of being British}} = \frac{0.04}{0.60}$$

$$= 0.067$$

The concept of marginal probability may be extended to multidimensional probability tables. For example, in three events X, Y and Z having outcomes x, y and z, the marginal probability of x is the probability of outcome x in event X, irrespective of the probabilities of y and z.

Marginals. These are the total univariate frequencies of two variates in a bivariate frequency table, shown separately in the marginal row and marginal column of the table. The term 'marginal' in this sense means 'at the margin' or 'at the end of the row or column' and must not be confused with the particular sense of 'differential', which the word has in economic analysis.

Example
The bivariate table below gives the age intervals of children vertically and height intervals horizontally. Each element of the table is a frequency value.

Age (Yrs)	Height (cm) Under 90	91 – 110	111 – 130	131 – 150	151 – 170	171 – 190	191 and over	Marginal frequency
11	1	—	—	—	—	—	—	1
12	1	7	1	—	—	—	—	9
13	—	1	6	12	4	2	—	25
14	—	1	12	20	3	4	—	40
15	—	1	7	7	12	4	—	31
16	—	—	—	2	7	1	—	10
17	—	—	—	—	—	1	2	3
Marginal frequency	2	10	26	41	26	12	2	119

The marginal column shows the frequency of all children of the ages stated, and the marginal row indicates the frequency of all children at the specific height intervals.

Markov chain.   A series of discontinuous states at intervals of time (t_1, t_2, ... t_p) such that the relative size of constituent classes or groups in a population at any given time is determined by stochastic probabilities inherent in the previous state, and not in any way affected by the earlier state(s) of the system. For example, in the illustration given under MARKOV PROCESS there is a certainty that all elements of the population which arrive at D, from whatsoever source, will return to A in the subsequent state. In the example each state represents a condition of the population after a given stochastic transition. A series of these states, each linked to the prior and subsequent states by transition probabilities is a Markov chain.

Markov process.   A process by which the probabilities of a series of discrete, discontinuous future states of a population are generated by existing states. A specific example of a Markov process is illustrated in the entry BRANCHING PROCESS where, given a collection of customers at time t, there is a probability that 0.5 of accounts outstanding will be paid at time $t + 1$, and that the other 0.5 of accounts will be paid later or remain unpaid.

The example illustrated is a specific case of a MARKOV CHAIN. More complex examples can be cited. In the example below, there are four classes, A, B, C and D. Let us assume that in the first state all constituents of a population are in A. In the second state 0.2 of A have moved to B, 0.5 have moved to C and the remaining 0.3 have moved to D. The constituents of B, C and D move to other groups in the following state as shown in the illustration. Unlike the more usual illustration shown under BRANCHING PROCESS, all constituents move from one group to another between one state and the next. The probability that a constituent of A will make the transitions C, B and A in three states is, for example, $(0.5 \times 0.1 \times 0.3) = 0.015$.

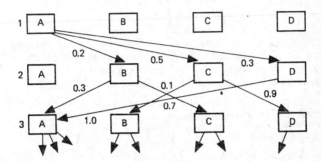

All transitions are repeated in the following states.

In matrix form the transitions can be represented:

	A	B	C	D
A	0.0	0.2	0.5	0.3
B	0.3	0.0	0.7	0.0
C	0.0	0.1	0.0	0.9
D	1.0	0.0	0.0	0.0

In some cases a steady state can be achieved, represented by a matrix which, when multiplied by the transition matrix, produces a matrix identical with itself.

Matched samples. If the elements in a pair (or set) of samples are such that each element in one sample corresponds to a particular element in the other on some characteristic other than the variable being investigated, the samples are said to be matched. The objective is to control the comparison in terms of some of the extraneous variables. For example, in a comparison of the hearing of industrial and clinical workers it might be desirable to match the samples in terms of age and sex.

Matrix. A rectangular arrangement of a group of elements which can be manipulated as though the table were one single element. A vector is a special case of a matrix where there is only one row (a row vector) or column (a column vector).

Examples are:

Matrix   Vectors

$$\begin{pmatrix} 5 & 3 & 1 \\ 8 & 9 & 2 \\ 7 & 6 & 8 \end{pmatrix} \qquad \begin{pmatrix} 1 \\ 3 \\ 8 \end{pmatrix} \quad \text{or} \quad (5 \quad 9 \quad 6)$$

A matrix which only consists of one single number or one single element $|6|$ is called a *scalar*, but the constituent units of a matrix (e.g. 5, 3, 1, etc. in the above case) are called ELEMENTS.

Some rules of manipulation of matrices are:

1. In order to *add* or *subtract* one matrix to or from another, simply add or subtract the *corresponding elements* of the matrices.

$$\begin{pmatrix} 5 & 4 \\ 3 & 8 \end{pmatrix} + \begin{pmatrix} 6 & 1 \\ 2 & 5 \end{pmatrix} = \begin{pmatrix} 11 & 5 \\ 5 & 13 \end{pmatrix}$$

but

$$\begin{pmatrix} 5 & 4 \\ 3 & 8 \end{pmatrix} - \begin{pmatrix} 6 & 1 \\ 2 & 5 \end{pmatrix} = \begin{pmatrix} -1 & 3 \\ 1 & 3 \end{pmatrix}$$

Addition and subtraction are only possible when all the matrices involved have the same dimensionality.

2. In order to *multiply* one matrix by another, aggregate the products of column 1 of the second matrix with row 1 elements of the first matrix, then column 2 with row 1, column 1 with row 2, and column 2 with row 2. Follow similar procedures for 3×3, 4×4 matrices etc. Note that matrices can only be multiplied by each other if the number of rows in the first matrix is identical with the number of columns in the second matrix.

Examples

$$\begin{pmatrix} 8 & 4 & 7 \\ 2 & 1 & 2 \end{pmatrix} \times \begin{pmatrix} 1 & 6 \\ 5 & 3 \\ 2 & 1 \end{pmatrix} = \begin{pmatrix} 42 & 67 \\ 11 & 17 \end{pmatrix}$$

but

$$\begin{pmatrix} 1 & 6 \\ 5 & 3 \\ 2 & 1 \end{pmatrix} \times \begin{pmatrix} 8 & 4 & 7 \\ 2 & 1 & 2 \end{pmatrix} = \begin{pmatrix} 20 & 10 & 19 \\ 46 & 23 & 41 \\ 18 & 9 & 16 \end{pmatrix}$$

Note that if the letters A and B are used to represent these matrices, AB ≠ BA, and the solution of

$$\begin{pmatrix} 8 & 4 & 7 \\ 2 & 1 & 2 \end{pmatrix} \times \begin{pmatrix} 1 & 6 \\ 5 & 3 \end{pmatrix}$$

is impossible.

3. Division is effected, when possible, by multiplying the matrix by the INVERSE of the divisor matrix. This is somewhat analogous to dividing a number by another, by multiplying the first number by the reciprocal of the divisor. (♦♦ DETERMINANT.)

Maximum likelihood estimators. A method of estimating the parameters of a population model from the sample data through the likelihood function. The maximum likelihood estimate of a parameter is a possible value of the parameter for which the likelihood function takes its largest value. It is the value of the parameter which is most plausible given the sample data.

Mean. The most commonly used AVERAGE or MEASURE OF CENTRAL TENDENCY or MEASURE OF LOCATION. The mean is computed by adding together the observations in a series and dividing by the number of observations in the series. In statistical symbols, the mean of a sample is indicated by the notation $\bar{x}$, and the parameter or population mean is indicated by the letter μ. Thus

$$\bar{x} = \frac{1}{n}\Sigma x_i \text{ and } \mu = \frac{1}{N}\Sigma X_i$$

where x_i and X_i are the sample and population observations respectively and where n and N are the sample and population sizes respectively.

Example
The values of x are 12, 17, 25, 28 and 33.
The total of values $(= \Sigma x_i)$ is 115.
The number of values $(= n)$ is 5.
The mean $(= \bar{x})$ is 115 ÷ 5 = 23.

Mean deviation. A measure of the spread or dispersion of a sample of population. Deviations of individual observations from a reference point, usually the arithmetic mean, are calculated. Negative signs are ignored so that only magnitudes of deviations are recorded and not their directions. These deviations are then averaged using an arithmetic mean. Thus if $x_1, x_2, \ldots x_n$ are observations and m is their mean (or reference point)

$$MD = \frac{1}{n} \sum_{i=1}^{n} |x_i - m|$$

For grouped data where a frequency f_i is associated with x_i

$$MD = \frac{1}{N} \sum_{i=1}^{k} f_i |x_i - m|$$

where $N = \sum_{i=1}^{k} f_i$.

Mean deviation, coefficient of.   MEAN DEVIATION expressed as a proportion of the MEAN. The coefficient is useful for comparing the mean deviations of distributions with dissimilar means.

Example
If the mean deviation of a variate is 23 and the mean is 50, the coefficient of mean deviation is 0.46.

Mean difference.   The average of the absolute differences of all possible pairs of values of a variate.

Example
Assume that a variate takes each of the values 5, 7, 12, 16, 19, and 25 with unit frequency. All absolute differences may be calculated, using the following table.

	5	7	12	16	19	25
5	0	2	7	11	14	20
7	2	0	5	9	12	18
12	7	5	0	4	7	13
16	11	9	4	0	3	9
19	14	12	7	3	0	6
25	20	18	13	9	6	0

The total of possible differences in the table is 280, and thus the mean difference is:

$$\Delta = \frac{280}{N^2}$$

$$= \frac{280}{36}$$

$$= 7.7\overline{7}$$

If we do not wish to include the comparison of each value with itself, we may define the mean difference as:

$$\Delta = \frac{280}{N(N-1)}$$
$$= \frac{280}{30}$$
$$= 9.\overline{3}$$

The calculation of mean difference is more complicated for variates with more complex frequency distributions. It is important in the study of CONCENTRATION and in the calculation of the GINI COEFFICIENT.

Mean range method (of estimating standard deviation). A rapid method of approximating the standard deviation of a population from the ranges of samples. It involves: (1) the calculation of the mean range of the samples; and (2) the multiplication of the mean range by a factor whose value is determined by sample size. These factors may be obtained from a statistical table, an extract of which is given below.

Example
A sample of 100 items is drawn from a population for the purpose of measuring their length. These are divided into ten subsamples of size 10, and the range of each subsample is calculated. Let us assume that the ranges of the ten subsamples are 7, 9, 9, 12, 6, 5, 8, 8, 10, 11, and employ R to denote the range. In this case the mean range is

$$\frac{\Sigma R}{n} = \frac{85}{10}$$
$$= 8.5$$

The mean range for all ten samples is then simply multiplied by a factor appropriate for the estimation of standard deviation where the sample size is 10, in this case 0.3249, so that an estimate of the standard deviation is

$$0.3249 \times 8.5 = 2.76165$$

Some useful values of the conversion factor for sample sizes are given below:

Sample size	Conversion factor	Sample size	Conversion factor
2	0.8862	7	0.3698
4	0.4857	8	0.3512
5	0.4299	10	0.3249

The method is a refined version of that used in SNEDECOR'S CHECK.

Mean square.   The arithmetic mean of the squares of differences between a set of given observed values of a variate and a particular value of the variate. In a restricted sense the variance of a set of values is a mean square, with particular reference to the mean of the observed values of the variate. The term mean square is more general and may be understood as the mean of squares of differences between observed values and any fixed value.

The term has a slightly different meaning in ANALYSIS OF VARIANCE. In compiling an analysis of variance table, the sums of squares for the variables are determined. The sums of squares are then divided by the appropriate number of degrees of freedom. Hence, in analysis of variance, mean square is

$$\frac{\text{Sum of squares}}{\text{Relevant degrees of freedom}}$$

Mean square deviation/mean square error. Both these terms are strictly understood in much the same sense as MEAN SQUARE, i.e. the average of the squares of differences between each of a set of observed values of a variate and a given value. If the given value used as a term of reference is the mean, then all three terms are identical with VARIANCE.

The mean square error is particularly relevant when dealing with biased estimators. Then

$$\text{Mean square error} = \text{variance} + (\text{bias})^2$$

Measure of central tendency, measure of location. The typical (or average) element in the population expressed as a single number or measure. This measure will be used for two purposes: (1) as a description of the population; (2) as a means of comparing two or more populations. There are a large number of possible measures we could use, and no single one of these measures is necessarily the 'best' measure. The appropriateness of a particular measure depends partly on the population being described and partly on the objective of the description. (◆◆ MEAN, MEDIAN, MIDRANGE, MODE.)

Measures, measurement. Both terms may be used to mean a standard common to a number of objects, by which they may be compared (e.g. a centimetre in measuring distance). The term measurement has more extensive uses, which include:
1. systems of measurement (e.g. pound or dollar for measuring money, or centigrade and absolute scales for measuring temperature); and
2. LEVELS OF MEASUREMENT (e.g. cardinal, ordinal, etc.).

Median. The middle value of a set of numbers arranged in order of magnitude. This is defined as the value below which (and above which) one half of the observations fall.

For example, the median of the nine numbers 1, 5, 7, 9, 11, 14, 17, 20, 22 is 11. The four numbers 1, 5, 7 and 9 are less than 11; the four numbers 14, 17, 20 and 22 are greater than 11.

Median test. A non-parametric test of similarity between two populations which relates the median of the two samples to the median of the combined sample.

Example
Observations of sample A are 1, 3, 9, 10, 15, 23, 34, 86, 90, 91, 99.
Observations of sample B are 1, 5, 7, 28, 46, 52, 91, 93, 97, 97, 99.
The median of the combined group is $\dfrac{34 + 46}{2} = 40$.

The test compares medians of the two populations A and B by examining the number of items in each of the samples ranking before and after the combined median. In this case seven items in Sample A and four in sample B fall before the combined median.

	<me	>me
A	7	4
B	4	7

A two-way table can now be constructed (as above) showing the numbers of observations in each of the samples which fall either below or above the median. If the total number of cases in both groups $(n_1 + n_2)$ is small we may now use FISHER'S TEST to test the null hypothesis, and if the number of cases is sufficiently large the CHI-SQUARED TEST with one degree of freedom may be used. (◆◆ YATES CORRECTION.)

Midrange. A measure of location or central tendency. It is the arithmetic mean of the largest and smallest observations in the data.

Example
Using the observations 5, 6, 12, 12, 15, midrange is $\frac{1}{2}(5 + 15) = 10$.

Minimum variance estimators. Often there are several possible estimators of a population characteristic which on the criterion of consistency or biasedness are equally desirable. However, the variances of their sampling distributions will differ. These are measures of the dispersion of possible values of each around the parameter value. Generally speaking, the smaller the sampling variance of the estimator the more concentrated is the distribution of its possible values around the parameter. An unbiased consistent estimator with a smaller variance will deviate less on average from the true parameter value. The one with the smallest variance is the minimum variance estimator and is usually to be preferred. In general, the term refers to the estimator with the smallest variance among those belonging to a particular class of estimators.

Missing plot/missing value.   In an experiment or a survey, some planned observations may be missing because of circumstances beyond the control of the researcher. These cases may be given a special CODE and special techniques are available for analysing data where some of the values are missing, but these are outside the scope of this book. The term 'plot' arises from the fact that most such early work was done in the area of agricultural experimentation where a missing plot was one for which, for some reason, no observation was available.

The use of plot as in plotting is completely different.

Mixed sampling.   A sampling plan is said to be mixed when it consists of a number of different sampling plans or techniques. For example, it may be decided to overcome the economic disadvantages of simple random sampling by using AREA SAMPLING or CLUSTER SAMPLING in one context and SYSTEMATIC SAMPLING in another.

Modal group.   Occasionally a set of numbers or observations are arranged and summarized as a grouped FREQUENCY DISTRIBUTION. The modal group is one with greater frequency for its breadth than either adjacent group. A frequency distribution can have more than one modal group. Modal groups can be defined for non-numeric nominal or ordinal scale frequency distributions.

Mode.   The value in a set of numbers which occurs more frequently than neighbouring possible values. There may be more than one mode. For example, the mode of ten numbers 1, 1, 2, 3, 3, 3, 7, 8, 9, 9 is 3; the fifteen numbers 2, 2, 2, 2, 3, 3, 5, 5, 5, 5, 6, 6, 7, 8, 9 have two modes, 2 and 5. The latter set of numbers is said to be BIMODAL. The normal distribution is UNIMODAL. It has one mode, which is coincident with its MEAN and MEDIAN. Asymmetric, or SKEWED DISTRIBUTIONS, also have single modes which do not coincide with their means and medians. In contrast, some frequency distributions are BIMODAL and MULTIMODAL, usually because the frequency distributions are themselves composites of two or more frequency distributions of different categories, which have differing modes.

A RECTANGULAR DISTRIBUTION has no mode.

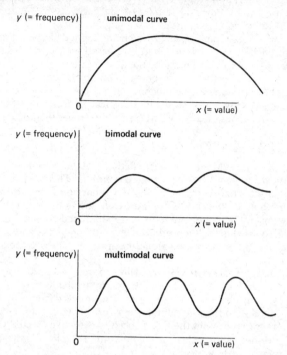

Model. A model is generally an attempt to summarize the complexity of the real world in the form of simplified statements or relationships. A model in this sense is not an exact replica of the system being studied – it does not attempt to describe every detail. Rather, it is a deliberate summary of the system and much of its usefulness arises from the fact that it concentrates on the broad or basic aspects of the system and discards what would otherwise be confusing detail. For example, an arterial road-map is useful because it ignores minor streets.

In statistics a model is generally expressed in terms of symbols (i.e. in mathematical form), usually as a set of equations. However, it is possible also to use diagrammatic models (♦ PATH DIAGRAMS), but for the purposes of estimation symbolic representation is usually necessary.

The word 'model' is now greatly overused and is often applied in situations where it is not really appropriate.

Model sampling.   The use of sampling methods which are based on broad assumptions about the way in which the variables of interest are distributed in the population, i.e. they are based on some implicit or explicit (but usually untested) model of the population. Among the methods in this class are haphazard sampling, purposive or judgement sampling and quota sampling, all of which are defined elsewhere. Most of the inferences we make individually are based on arbitrary samples of this kind – judging a restaurant by the quality of a single meal we eat there; evaluating a lecturer by listening to one or two of his lectures; judging an acquaintance by his behaviour on one or two occasions when we meet. However, for sound statistical inferences we should base our observations on a PROBABILITY SAMPLE from the population.

Moment.   ◗ FIRST MOMENT.

Monte Carlo method/Monte Carlo simulation.   A method of solving a mathematical problem by using repeated sampling. The most common way of doing so is to construct a model and generate samples from the model, and then use the results obtained from the samples to estimate the behaviour of the model. Sometimes the simulated results are compared with the real population process which the model is assumed to represent, and differences are studied in order to improve the simulated process.

The stochastic process is generated in different ways. In business games the probability of a market becoming available may be represented by the probability of drawing a particular card from a pack. In computer simulation random numbers may be introduced into a computer program to represent the risk associated with an expected value. Monte Carlo simulation is also used to determine empirically the properties of complex estimators.

Moving average.   When data occur as a time series groups of adjacent or overlapping observations can be averaged to smooth out short-run fluctuations. If $x_1, x_2, x_3, x_4, \ldots$ are successive values of the series a three point moving average would be $\frac{1}{3}(x_1 + x_2 + x_3)$, $\frac{1}{3}(x_2 + x_3 + x_4)$, $\ldots$ and these values would be associated with time periods 2, 3, $\ldots$ respectively. Unequal weights (adding up to 1) can be used as, for example, in the Spencer fifteen-point moving average.

Multimodal.   Having many MODES. Frequency distribution curves may consist of many modes when they are the composites of a number of distribution curves. For example, a department store may consist of a number of departments (groceries, hardware, electrical goods, etc.). The modal account values of frequency distributions of each of the departments would be different. An uncategorized frequency distribution consisting of all values of accounts in a given period might be multimodal.

Multiphase sampling.   It is sometimes convenient and economical to collect certain items of information on the whole of the units of a sample and other items of information (usually more intensive or detailed) only for a subsample of the units. This is called two-phase sampling. Further phases may be added as required. All such sampling schemes are called multiphase.

In carrying out an investigation of absenteeism within a country, for example, a large sample of firms may be selected in the first instance (the first phase) and some simple descriptive information may be collected about each of the firms in order to give a general idea of the population characteristics. Using the information collected in the first phase, a much smaller sample of firms (a subsample of the original sample) may be selected for an intensive investigation (the second phase). A further subsample (the third phase) might involve only one or two firms in which interviews with the workers might be carried out. This illustrates a useful feature of multiphase sampling – the information from earlier phases may be used to improve the selection of the later phase subsamples. In general, multiphase sampling is only desirable if the cost of data collection for the first phase is considerably lower per unit than for the second phase. Thus the information for the first phase is generally collected using different and much cheaper methods – for example, a simple postal questionnaire or the examination of existing records. Multiphase sampling should not be confused with MULTISTAGE SAMPLING.

In contexts other than survey sampling, the term multiphase sampling may be used to refer to SEQUENTIAL SAMPLING.

Multiple correlation/multiple correlation coefficient.   The term multiple correlation means the correlation between a single variable and a set of variables taken jointly. In MULTIPLE REGRESSION it is the correlation between the estimated values of the dependent variable and the observed values.

The multiple correlation coefficient is usually denoted by R as distinct from ρ (the correlation coefficient between any two variables or between the dependent variable and any one of the regressors).

Example
It may be computed that a set of observations can best be explained by the regression equation

$$\hat{Y} = 20 + 3X_1 - 5X_2$$

Let us assume that this model produces the following estimates and that these are compared with the observations themselves.

Observations	Y	2	6	8	11	12	15	18
Estimates (using the above model)	$\hat{Y}$	4	7	8	12	14	15	12

The multiple correlation coefficient (i.e. that between observations and estimates of Y) helps us to assess whether the regression model is a useful estimator of Y.

Note that, just as ρ^2 is the proportion of the variance of cither variable explained by its regression on the other, so is R^2 the proportion of the variance of the dependent variable explained by the explanatory variables in the MULTIPLE REGRESSION.

Multiple regression. The regression of one dependent variable (REGRESSAND) on more than one independent variable (REGRESSORS). It is an extension of simple linear regression and the equation is of the form

$$y = \beta_0 + \beta_1 x_1 + \beta_2 x_2 + \beta_3 x_3 \ldots \beta_k x_k + \varepsilon$$

The regression coefficients $\{\beta_i\}$ are estimated simultaneously. Many computer programs exist which carry out the operation of estimation. (◗ REGRESSION ANALYSIS.)

Multiple sampling. Multiple (or multiphase) sampling is a type of sample design in which some information is collected on the whole sample (usually a large sample), and further information is collected only for subsamples of the whole sample. When there is only one such subsample the design is called TWO-PHASE SAMPLING. ◗ DOUBLE SAMPLING.

Multiple stratification. When a sample is stratified by two or more factors, it is multiply-stratified. If we wish to stratify industrial organizations by size and region, for example, we must set up a stratum corresponding to each cell of the cross-classification of size by region. If we have five size categories (A, B, C, D, E) and four regions (N, S, E, W) we will need $5 \times 4 = 20$ strata in all. (◗ STRATIFICATION.)

Multiplication principle (in probability). This principle states that: (1) if two outcomes are independent of each other, the probability of both outcomes occurring is the product of the probabilities of each of the outcomes; and (2) if the two outcomes are such that one outcome is dependent on the other, the probability of both outcomes occurring is the probability of the first event multiplied by that of the second outcome given the first.

Examples

1. If two dice are cast, the probability of obtaining a '6' from die A, $P(A)$, is independent of that of obtaining a '6' from die B, $P(B)$. Let $P(A \cap B)$ be the probability of two '6's, one from each die. In this case the special condition of (1) above is applicable. $P(A \cap B) = P(A) . P(B)$, and thus, as in this particular example, both $P(A)$ and $P(B)$ are 1/6, $P(A \cap B)$, the probability that two '6's will be cast by throwing both dice is calculated

$$P(A \cap B) = \frac{1}{6} \times \frac{1}{6}$$

$$= \frac{1}{36}$$

2. If an urn contains two red balls, one blue ball and three black balls, the probability of obtaining a particular colour on the second draw (B) is dependent on that on the first draw (A). In this case $P(A \cap B) = P(A) . P(B|A)$, where $P(B|A)$ is the probability of a particular colour on the second draw, given the first draw. Thus the probability of obtaining black balls on each occasion is calculated in the following way

$$P(A) = \frac{3}{6}$$

but $P(B|A)$ is 2/5 since if $P(A)$ occurs only two of the remaining balls will be black. Thus

$$P(A \cap B) = \frac{3}{6} \times \frac{2}{5}$$

$$= \frac{6}{30}$$

$$= \frac{1}{5}$$

Multistage sampling.   Although cost considerations frequently rule out the practicability of an element sample from a human population, a CLUSTER SAMPLE in which whole groups of elements are selected together for observation may be considered unsatisfactory because of its low precision. Consequently a compromise solution may be obtained by subsampling from within each of the selected clusters. The units used in the subsampling (the second stage of selection) may themselves be smaller clusters of elements, and a third or fourth stage may be carried out before the designated elements are obtained. This is multistage sampling, and a high proportion of the sample designs used in sample surveys are of this kind.

Consider the situation where we wish to predict the result of a general election by interviewing a sample of voters in the country. We might expect reasonably good precision by interviewing a simple random sample of 1000 people. If we used simple random sampling, the respondents would probably be very widely dispersed over the whole country. The travel costs involved in the fieldwork would be enormous, since, for example, if only one respondent happened to live in Orkney we would have to send an interviewer all the way there to obtain that interview. On the other hand, if we were to select, say, two polling districts and interview all the individuals within them we would have very little confidence in our results. This conflict between cost and precision might lead us to use multistage sampling. As a first stage we might randomly select fifty constituencies from the 635. The constituencies would be the primary sampling units. Secondly, within each of the selected constituencies we might then randomly select a number of polling districts – the polling districts would be the secondary sampling units. Finally, as a third stage we might select a simple random sample of individuals within each of the selected polling districts. Thus the final-stage units are the elements. This overall strategy – multistage sampling – provides a compromise solution. The selected sample would not be as costly as a simple random sample of elements nor would it be as imprecise as a cluster sample.

In selecting a sample from the Register of Electors in Great Britain it would be possible to select a simple random sample of elements. In many situations an element sample is simply not possible. In the United States, for example, no centralized list similar to the Register of Electors exists. In such a situation, multistage sampling has a considerable additional advantage. Initially all that is required is a list of the primary sampling units. A list of the secondary sampling units is required only for the selected psu's. Similarly a list of the third-stage units (perhaps the elements) is required only for the selected secondary stage units. Thus, in effect, multistage sampling may make it possible to select a sample where an element sample would be impossible. (◆◆ DESIGN EFFECT, SAMPLING ERROR.)

Mutually exclusive.   Two events are mutually exclusive if the occurrence of one of the events renders the occurrence of the other event contemporaneously imposible. Thus it is not possible for a person to throw a single coin and obtain both a head and a tail at the same time, nor is it possible for a candidate both to pass and fail the same examination at the same time.

Let us assume that a position is advertised and that there will certainly be an applicant for it. This certainty is expressed mathematically as $P = 1$ (i.e. probability = 1). Applicants may be either male or female, and thus the one event of a male being the first applicant for the position entirely excludes that of a female being the first applicant. If the events are mutually exclusive and exhaustive and the probability of a male applicant is p, then that of a female applicant is $1 - p$, and their sum is 1.

The rule applies to all mutually exclusive probabilities. The probability of either event X or event Y occuring, if X and Y are mutually exclusive, is $P(X) + P(Y)$. (◗ ADDITION PRINCIPLE.)

Mutually exclusive classification.   The division of a population, sample or set of observations into a set of classes or categories which are mutually exclusive. The classes may either be ATTRIBUTE categories (e.g. sex, race, occupation, nationality) or variate interval classes (1–10, 11–20, 21–30, 31–40). The classification may either be UNIVARIATE (as represented in a HISTOGRAM or other single distribution of mutually exclusive categories); or it may be bivariate (e.g. the $m \times n$ table of mutually exclusive categories required for chi-squared or modified chi-squared tests); or it may be multivariate.

N

Napieran logarithms. The original tabulation of logarithms, devised by John Napier (1550–1617) the Scottish mathematician after whom they are named. Napieran (natural) logarithms are based on the mathematical quantity e (≈ 2.718) which is the sum of the series

$$e = \frac{1}{0!} + \frac{1}{1!} + \frac{1}{2!} + \frac{1}{3!} + \frac{1}{4!} + \frac{1}{5!} + \ldots$$

The interest in the entity e is that if a quantity P grows at a rate rt where r is a constant and t is the time elapsed its value V at any stage may be written

$$V = Pe^{rt/100}$$

and by transposition $V/P = e^{rt/100}$ and thus $rt/100$ is the natural logarithm of V/P to base e. Apart from the invaluable use of Napieran logarithms, as for example in compiling compound interest tables, the quantity e is so important in calculations that Napieran logarithms have continued to be used despite the subsequent production of decimal logarithms.

The main problem which Napieran logarithms present is their lack of simplicity because of the non-integer nature of e. The difficulty was subsequently overcome by another mathematician named Briggs, who devised a table of logarithms to the base of 10. ($\blacklozenge$ LOGARITHM, e (EXPONENTIAL).)

Nested classification. If one of the factors in a two-way classification is such that no MAIN EFFECT can be attributed to it, then that factor is said to be nested within the main classification.

Example
If a company has four breweries and employs three different inspectors in each, the factor 'inspector' is said to be nested within the main classification 'brewery'. This is because in the table below the label Inspector 1 describes a different inspector in each of the four breweries.

inspector \ brewery	1	2	3	4
1				
2				
3				

(◆◆ CROSSED CLASSIFICATION.)

Nested sampling. A term sometimes used as an alternative to the more widely used MULTISTAGE SAMPLING.

Net error. That part of the total or GROSS ERROR which does not cancel out when cases or observations are averaged or totalled. The examination of net error is usually easier than the examination of gross error and can be justified by the argument that it is the net error which introduces a BIAS into the results of the survey or experiment when these results are expressed, as is often the case, in terms of totals or means. However, not only are measures of association affected by gross errors, but the study of net errors alone will not lead to the elimination or reduction of total error. (◆◆ RESPONSE ERRORS.)

Neyman allocation. A name given to OPTIMAL ALLOCATION of a sample when costs are assumed to be equal in each stratum. This optimal allocation was first derived by Neyman in 1934.

Neyman–Pearson lemma. The Neyman–Pearson lemma provides a methodology for finding the best test among those with the same probability of type I error, i.e. the test with the highest power (the lowest probability of type II error).

The test is based on the LIKELIHOOD RATIO and an example is given in the entry LIKELIHOOD RATIO TEST.

Nominal scale. The simplest form of measurement is on a nominal scale. Here, individuals or elements are allocated to categories of the characteristic being measured where no relationship is assumed to exist between the categories. Thus if numbers are attached to (associated with) the categories, these numbers are labels and have no other meaning. The competitors in a dancing competition may be labelled in this way. A nominal scale is simply a classification or categorization. With data of this kind, the only statistical methods of analysis we can use are those based on frequencies or counts – the mode, contingency tables, frequency distributions, etc. Nominal scales are sometimes called associative scales, or scales of associative measurement.

Nomogram. A form of multiple line diagram in which the scales for different functions of a variable are drawn alongside each other and corresponding values are in line with each other. Nomograms facilitate the identification of differences between scales, but are not as accurate as tables.

Example

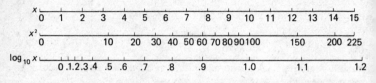

Non-coverage.   Failure to include units or sections of the SURVEY POPULATION in the operational SAMPLING FRAME. For example, if we select a sample of adults using the Register of Electors as a sampling frame a certain proportion of the survey population (defined as consisting of all adults living in Great Britain, say) will be excluded from consideration since their names do not appear on the Register. We must distinguish between this situation and the deliberate exclusion of sections of a larger population from the survey population. For example, many surveys of adults are confined to those living in households – residents of institutions are often excluded for practical reasons. Non-coverage refers only to the exclusion of units from consideration due (1) to faults in the lists themselves or (2) to failures of the survey procedures – thus faulty execution of fieldwork operations may also lead to non-coverage. It is important to bear in mind that the extent of non-coverage can only be estimated using information from outside the survey itself. A comparison of an estimate based on the sample itself with a reliable, current external source may provide some information on the extent of the problem. To decrease the impact of the non-coverage, a special quality check may be made in part of the sample using different, more thorough, and possibly more expensive, fieldwork procedures. The sample estimates may be adjusted in the light of the findings. Alternatively some linking procedure such as the HALF-OPEN INTERVAL may be used to increase the coverage of unlisted units.

Non-determination, coefficient of (k^2).   A measure of alienation, which is the square of the COEFFICIENT OF ALIENATION, and related to the PRODUCT-MOMENT CORRELATION coefficient (ρ).

$$k^2 = 1 - \rho^2$$

The coefficient measures the proximity of relationship between two variates to zero-correlation.

Non-parametric methods.   ◗ DISTRIBUTION-FREE METHODS.

Non-response.   The failure to obtain the required information from members of the selected sample for any reason. In order to measure the rate of non-response it is necessary to keep careful and complete records of the selection of individuals and the outcome of the field-work. The non-response rate can be defined as the proportion of designated individuals for whom the survey information was not obtained. However, the sources of non-response are so diverse that it is usually desirable not to use a single gross rate but to split it up according to the reason for non-response. Some failures to obtain a response may be due to the death, incapacity or absence of the designated individual at the time of the survey. Others may be due to refusal to cooperate despite being contacted. In some cases it may be too difficult or too expensive to find the designated individual – a lighthouse keeper, for example. It is also possible, although one hopes not common, that completed schedules or questionnaires may be lost in the post or accidentally destroyed. These also should be classified as non-response since although the information has been collected it never reaches the stage of being analysed and therefore does not contribute to the survey results.

All the classes of non-response refer to eligible respondents, i.e. those who belong to the survey population. If individuals are selected in error (for example, a fifty-year-old in a survey of people under the age of forty) then the failure to obtain information does not constitute non-response. In fact, including such information would itself be a cause of error. The response rate should be calculated on the basis of responses and non-responses among eligible population elements only.

The effect of non-response on the survey results will vary depending on the source of the non-response and also the purpose and nature of the survey. Refusals will typically be a more serious source of error than unavailability since those who refuse will often differ in many respects from those who agree to participate. However, if we are carrying out a survey of, say, leisure activities outside the home, those who are absent from home when the interviewer calls may well have a different pattern of behaviour to the rest of the population and failure to obtain information from them may cause serious biases in the results. Thus the classification and seriousness of different sources of non-response will vary from one survey to another. The dominant sources should be classified separately for each survey.

The terms partial non-response or incomplete response are used to describe cases where some, but not all, the information has been obtained from an individual.

Non-sampling errors.   Errors in estimates which are not due to the fact that the information is based on a sample from the population. Thus non-sampling errors are important even when information is obtained for the whole population, as in the case of the Census of Population. A useful distinction can be made between errors of observation and errors of non-observation. Observation errors are partly in the domain of the social statistician but are also fundamental to the whole issue of scientific measurement. They affect complete population enumerations as well as sample surveys as, for example, the problems of measuring attitudes to race, religion or violence. Some problems specific to sample surveys are discussed under RESPONSE ERRORS. Errors of non-observation arise from failure to obtain data from part of the population and the two components are NON-COVERAGE, which is a failure to include part of the population in the operational list of population elements, and NON-RESPONSE, which means a failure to obtain the appropriate data from designated elements, i.e. those from whom it was decided to collect information.

Normal curve, normal distribution.   The most commonly used type of continuous frequency distribution (or distribution curve). It coincides with the distribution of many observations of natural phenomena (such as height, weight and ability), and is also the limiting form of the BINOMIAL DISTRIBUTION where $p = q$. The curve is BELL-SHAPED and symmetrical.

The measurement of the height of the curve dF depicts frequency (or relative frequency if the total area is 1) and is given by the equation

$$\mathrm{d}F = \frac{1}{\sigma\sqrt{(2\pi)}}\ e^{-\frac{1}{2}\left(\frac{X-\mu}{\sigma}\right)^2} \qquad \mathrm{d}X,\ -\infty \leqslant X \leqslant \infty$$

If X is a random variable, having binomial distribution with the parameters n and p, then the distribution of the variable

$$Z = \frac{X - np}{\sqrt{(npq)}}$$

tends to *standard Normal* $(0, 1)$ as n tends to infinity.

We may write the height, dF as

$$\mathrm{d}F = \frac{1}{\sqrt{(2\pi)}}\ e^{-\frac{1}{2}Z^2} \qquad \mathrm{d}Z,\ -\infty \leqslant Z \leqslant \infty$$

where $Z = (X - \mu)/\sigma$.

It is thus possible to measure distance from the mean of a normal curve, not merely in absolute measurements, which may change from one case to another, but in terms of the standard NORMAL DISTRIBUTION. Its points of inflexion (i.e. from convex to concave) occur one standard deviation on each side of the mean. The 'Z-score' measures the area under the curve in standard deviations along the horizontal axis from the mean. Some commonly used values are:

Standard score (Z)	Tail area $(1-\varphi(Z))$	Standard score (Z)	Tail area $(1-\varphi(Z))$
1.000	0.1587	2.500	0.0062
1.500	0.0668	3.000	0.0013
2.000	0.0228	3.500	0.0002

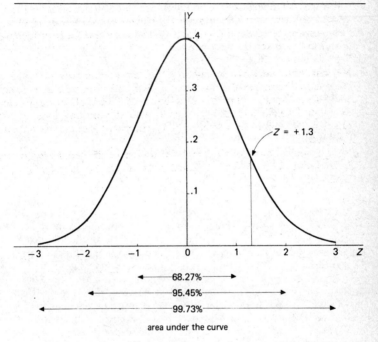

area under the curve

The curve and associated distribution are often called Gaussian and Laplacean, because of their association with Gauss and Laplace.

Normal equations. A set of simultaneous equations derived in the application of the LEAST SQUARES METHOD, for example in multiple regression analysis. They are used to estimate the parameters of the model.

Normalization. The conversion of a set of raw values or raw scores into a scale which corresponds to a known frequency distribution. This need not be the Normal symmetrical frequency distribution, but conversion to Standard Normal score equivalents is more commonly used than conversion to scores approximating other relevant frequency distributions. The process is sometimes assisted by using Normal Probability paper, i.e. paper ruled with a set of bunched central graph lines, so that the normal probability curve is plotted as a straight line. Normalized scores are sometimes termed Standard Normal scores, or z-SCORES.

Not applicable/does not apply. A classification used to distinguish the situation where a question is not applicable to a particular respondent from a refusal to answer the question. The question, 'How many years did your first marriage last?' is not appropriate to those who have not been married. To record the answer zero (no years) or 'Refused to answer' would be misleading in the analysis.

Not asked. In most surveys, no matter how well conducted, information on some items will be missing for some of the individuals who have otherwise supplied all the requested information. It is important to distinguish between three possible causes: (1) where the respondent refused to supply the answer (refusal); (2) where the question was not applicable to that respondent (not applicable); (3) where the interviewer, through an oversight or for some other reason, neglected to ask the question. The last should be coded separately in the data as 'Not asked'.

Not-at-homes. One of the categories of NON-RESPONSE. It includes those who are away from home throughout the period of the survey and those who are simply out at the time of the call. In the latter case, calling back at another time may be sufficient to obtain an interview. The nature of the population being studied is important. People in full-time employment are not normally at home during the day. Young single people are more frequently not at home in the evenings than parents with young children. Those who live in rural areas spend more time at home than urban dwellers. Thus the *time of calls* is important. The *season of the year* may also affect the response rate – holiday periods are not favourable for interview surveys. It is also worth distinguishing between 'Nobody at home' and 'Respondent absent' situations. In the latter case it may be possible to find out when the respondent may be found at home and sometimes an appointment for the interview may be arranged.

Not reached/Not found. One of the categories of NON-RESPONSE. Failure to contact an individual may be caused by inaccessibility or by other factors beyond the control of the interviewer. This category should not be large in most surveys although it may be an important component of total non-response in some mail/postal surveys or in PANEL STUDIES, which involve following up movers.

Null hypothesis.   A particular hypothesis (usually indicated by the symbol H_0) being tested, as distinguished from any alternative hypotheses that may be considered in the context.

Any hypothesis about the value of an unknown parameter (e.g. $\pi = 0.5$ or $\beta_1 \neq 1.75$) may be regarded as a null hypothesis in a particular context. However, in statistical usage, the term often means a hypothesis that there is no difference (e.g. between the sample mean and the mean of a parent group or between the means of two samples). The null hypothesis may be rejected at a given significance level. (◗ SIGNIFICANCE TESTING.)

O

Observation. The term has two distinct meanings: (1) the observed value of a variate for an element is called an observation; and (2) observation is the process of obtaining information directly, for example, by observing behaviour – as distinct from obtaining information from reports or records – by asking people about their behaviour, for instance.

Ogive. A continuous form of the cumulative frequency curve, which assumes a very large number of observations. If the cumulative frequencies are plotted against the ordinate axis and the variate values are plotted against the abscissa, any unimodal cumulative frequency distribution will have an S-shaped appearance, taking the form illustrated below.

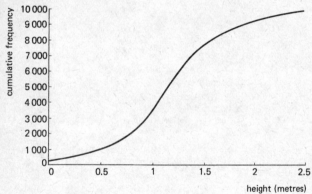

An ogive showing the measurements of people under and including a given height

The name 'ogive' was coined by Galton from the ogive style used in architecture.

Open question, open-ended question. In an open question it is the respondent who decides the form, length and content of the answer.

The interviewer must record as much of the answer as possible. An example of an open-ended question might be, 'What are the things you like about working here?' In this case the respondent is given freedom to decide which aspects of his job to specify in his answer – e.g. the job itself, his fellow-workers, his supervisors, management, working conditions or even travel to work – and indeed may mention all, some or none of these things. The advantages of open-ended questions are: (1) they allow the respondent to mention what is important to him; and (2) since the response is recorded in full by the interviewer, CODING can be carried out uniformly in the office rather than individually by the interviewer. The disadvantages are: (1) the responses to open-ended questions are most sensitive to influence by the interviewer and may lead to interviewer bias or INTERVIEWER VARIANCE; (2) the coding operation is much more difficult; and (3) the interviewer may have difficulty in writing down the full response which may lead to biases in the selection of which part of the response to record.

Optimal allocation.   Decisions regarding the allocation of the sample arise in two situations. The first is in STRATIFIED SAMPLING where the allocation of the sample between the strata must be determined. The second is in the case of MULTISTAGE SAMPLING, where the number of elements to be subsampled from each primary sampling unit must be decided.

In stratified sampling the simplest procedure is PROPORTIONAL ALLOCATION where the same proportion of the elements is selected within each stratum. However, two factors may influence our decision. First, the cost of collecting the information may differ from one stratum to another. Second, the variability (dispersion) of the elements may be different for each stratum.

The optimal (most efficient) method of allocation is to choose proportionately fewer elements from the more expensive strata and to choose proportionately more elements from the more variable strata. As a simple example consider a population with the following structure; here the total size n is fixed and the cost of an observation is the same in both strata.

N (population size) = 150 000
N_1 (size of stratum 1) = 75 000
N_2 (size of stratum 2) = 75 000
n (total sample size) = 600
S_1 (standard deviation in stratum 1) = 4
S_2 (standard deviation in stratum 2) = 2
c_1 (cost of one interview in stratum 1) = 1
c_2 (cost of one interview in stratum 2) = 1

The general formula by which we calculate the allocation of the two strata is

$$n_h = \frac{N_h S_h / \sqrt{c_h}}{\displaystyle\sum_{h=1}^{2} N_h S_h / \sqrt{c_h}} \times n$$

where h denotes the stratum number and can, in this case, take values 1 or 2 (corresponding to strata 1 and 2).

In the example $c_1 = c_2 = 1$ and hence the formula can be written as

$$n_h = \frac{N_h S_h}{\Sigma N_h S_h} \times n$$

Therefore, n_1, the sample size in stratum 1, can be calculated as:

$$n_1 = \frac{N_1 S_1}{N_1 S_1 + N_2 S_2} \times n$$

and n_2, the sample size in stratum 2, can be calculated as:

$$n_2 = \frac{N_2 S_2}{N_1 S_1 + N_2 S_2} \times n$$

Let us first calculate the denominator which is the same in both cases.

$$
\begin{aligned}
N_1 S_1 + N_2 S_2 &= (75\,000)\,(4) + (75\,000)\,(2) \\
&= 300\,000 + 150\,000 \\
&= 450\,000
\end{aligned}
$$

Thus

$$n_1 = \frac{(75\,000)\,(4)}{450\,000} \times 600 = \frac{300\,000}{450\,000} \times 600 \qquad (1)$$

$$= 400$$

Similarly

$$n_2 = \frac{(75\,000)\,(2)}{450\,000} \times 600 = \frac{150\,000}{450\,000} \times 600 \qquad (2)$$

$$= 200$$

Thus the total sample of $n = 600$ is allocated as 400 from stratum 1 and 200 from stratum 2. The higher proportion selected from stratum 1 is due to the fact that the variability is greater in stratum 1. If proportional allocation had been used, half of the sample would have been selected from each stratum; i.e. $n_1 = n_2 = 300$.

However, the allocation given by (1) and (2) will lead to a much more precise estimator. It is important to note that this gain in precision will occur only if our information about the variability of the elements (S_1 and S_2) is correct. If this information is seriously wrong, the consequence may be a very imprecise estimator. Thus optimal allocation should only be used when we have some confidence in the prior information about the strata.

Ordinal scale.   A SCALE OF MEASUREMENT where the scores allocated to the individuals represent a ranking or ordering of the characteristic being studied. In attitude measurement, ordinal scales are commonly used. Thus individuals could be classified as Very authoritarian, Authoritarian, Fairly authoritarian or Not at all authoritarian. These four categories could be allocated the scores 4, 3, 2 and 1 respectively. The numbers in this case represent the ranking of the individuals in terms of authoritarianism but could be replaced by any set of numbers which are in the same order (e.g. 107, 54, 32, 19) without losing any information. Thus scores on ordinal scales should not be treated as cardinal numbers and may not, for instance, be added to or subtracted from one another in any meaningful way.

In addition to the statistical methods available for use with nominal scale data, the class of statistical techniques known as NON-PARAMETRIC METHODS can be applied to ordinal data.

Ordinate.   The vertical axis of a graph, the corresponding axis (or base line) being known as the ABSCISSA, or sometimes as the co-ordinate axis.

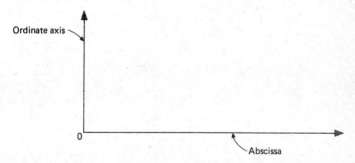

Convention and statistical usage employ the ordinate scale:
1. for frequency measurements in a frequency chart;
2. for cumulative frequency measurements in the case of an ogive;
3. for measurements of a dependent variable in a regression chart;
4. for ratio–scale measurements in a semilogarithmic chart;
5. for value measurements in the case of historigrams and time-series charts where the abscissa is usually reserved for measurements of time.

Orthogonality.   The original meaning of the word was 'right-angled', but because of different usages in matrix algebra and statistics the word has come to have several distinctly different applications. For example, an orthogonal transformation is a method of linear substitution for transforming one set of rectangular co-ordinates into another, so that the sum of squares of the variables is left unchanged. A matrix is said to be orthogonal if its transpose is equal to its inverse. In vector analysis the orthogonal complement to a vector is the set of all vectors perpendicular to it, and two vectors are orthogonal to each other if the scalar product on multiplying one by the other is zero.

In experimental design, the term orthogonal describes a condition where cell frequencies are constant.

Overcoverage.   Including elements for consideration which do not belong to the defined SURVEY POPULATION. It is not normally as serious a problem as NON-COVERAGE.

P

Paasche index. This is a form of index number that applies current period weights to the items that make up the index. For example, in a price index the value of goods currently purchased is compared to the same quantities of goods revalued at prices of different periods.

If $q_i^{(1)}$ are quantities of the component goods currently being purchased $p_i^{(1)}$ and if $p_i^{(0)}$ $i = 1, 2, \ldots n$ are the prices pertaining to a different period, then the Paasche index is given by

$$\frac{\sum_{i=1}^{n} p_i^{(1)} q_i^{(1)}}{\sum_{i=1}^{n} p_i^{(0)} q_i^{(1)}}$$

It tends to give a lower value than the Laspeyres index because people tend to shift their purchasing pattern away from commodities whose relative price increase is large and towards those whose relative price increase is small.

Paired comparison. The term is used in two contexts:
1. Difficulties arise in asking people to rank a large number of objects in order of preference. As an alternative the objects are presented to the respondent in pairs and he is asked to state which of the two he prefers. This is the method of paired comparisons and is widely used in testing taste preferences, for example. The method can produce inconsistencies – for example, a respondent may say he prefers A to B, B to C, but C to A. This is a CIRCULAR TRIAD. The consistency (or inconsistency) of paired comparisons is measured by the COEFFICIENT OF CONSISTENCE.
2. In a more general sense, the term is used to describe a comparison of two samples of equal size where it is possible to pair off the members of one sample against the members of the other.

Paired selection sample design. A sample design in which two units are selected from each stratum. It is a form of REPLICATED SAMPLING and has two particular advantages. First, by permitting stratification to be carried to the stage where there are only two selections per stratum, it allows the maximum amount of stratification consistent with estimating the SAMPLING ERRORS from the sample data. Second, the method permits the simplest formulae to be used for the computation of sampling errors.

Paired selections. The name given to two units selected randomly from within a stratum. Paired selections form the basis of a PAIRED SELECTION SAMPLE DESIGN.

Parameter. The value of a particular measure (e.g. mean, standard deviation) of a population or parent group.
For example, in

$$f(x) = \frac{1}{\sigma\sqrt{2\pi}} \; e^{-\frac{1}{2}[(x - \mu)/\sigma]^2}$$

μ and σ are parameters. Parameters are estimated, or hypotheses are tested concerning them, on the basis of the statistics obtained from samples taken from the population(s).

Parametric. That pertaining to a parameter. For example, a parametric hypothesis is a hypothesis relating to the parameters of a given distribution.

Parity. Equality: the term 'parity' is more commonly used in expressing equality of two sets of terms in an equation.

Partial association. Association between two variables when the third is controlled or at a fixed level. The term is analogous to PARTIAL CORRELATION.

Partial correlation. The correlation between two variates, controlling for one or more other variates. For example, if sales of ice-cream and sales of beachwear are highly correlated because of the mutual correlation with a third variable, temperature, partial correlation would seek to discover the extent to which they are correlated, controlling for temperature measurements. If the two sales variables are expressed as Y_1 and Y_2 and the temperature variable is expressed as X, partial correlation first seeks to determine the least-squares regression equations $Y_1 = a_1 + b_1X$ and $Y_2 = a_2 + b_2X$. Then sets of estimates of Y are produced using each of these equations and the deviations between observed values and both sets of estimates are calculated. The two sets of deviations are then correlated with each other, and the resulting correlation coefficient expresses the extent of partial correlation.
In practice, the following simple formula is usually employed to obtain the partial correlation coefficient between Y_1 and Y_2 controlling

for X in terms of the simple correlation coefficients between the three variables in turn:

$$r(Y_1Y_2 . X) = \frac{r(Y_1Y_2) - r(Y_1X)r(Y_2X)}{\sqrt{\{[1 - r(Y_1X)^2][1 - r(Y_2X)^2]\}}}$$

where $r(Y_1Y_2.X)$ is the partial correlation coefficient between Y_1 and Y_2 given X; and $r(Y_1Y_2)$, $r(Y_1X)$ and $r(Y_2X)$ are the simple correlation coefficients between Y_1 and Y_2, Y_1 and X, and Y_2 and X respectively.

Pascal triangle. A triangle of numbers named after Pascal (1623–62), to whom it is attributed, showing the coefficients of the binomial expression raised to the power of k ($k = 1, \ldots n$, where n is the number of rows of the triangle). Assuming the outcomes to be head (H) or tail (T) on tossing an unbiased coin, and k to be the number of occasions on which the coin is tossed, the Pascal triangle is illustrated below.

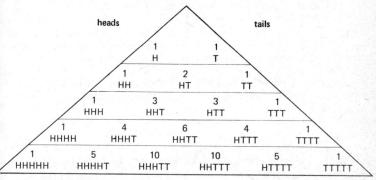

The triangle gives the coefficients where the binomial expression ($h + t$) expanded k times (i.e. $(h + t)^3 = h^3 + 3h^2t + 3ht^2 + t^3$, etc.). Thus, if a coin is tossed three times, the likelihood of outcomes is measured by dividing the respective coefficients (or row elements) by the row total. The likelihood that all outcomes will be heads is ⅛, that there will be two heads and one tail is ⅜, that there will be one head and two tails is ⅜ and that all outcomes will be tails is ⅛. Each element of each row is the sum of adjacent elements from the previous row, e.g. $1 + 3 = 4$.

Path analysis. A method by which the coefficients of a PATH MODEL are estimated. Consider the following simple path model:

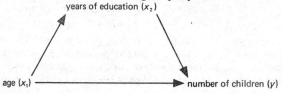

The equations can be written as:

$$x_2 = p_{21} x_1 \tag{1}$$
$$y = p_1 x_1 + p_2 x_2 \tag{2}$$

indicating that x_2 (education) is determined by x_1 (age) and that y (number of children) is determined both by x_1 (age) and x_2 (education). However, the relationship is not exact – other factors not included in the model will also affect the variables – and residual or error terms should be added to the equations (1) and (2), giving:

$$x_2 = p_{21} x_1 + p_{2u} x_u \tag{1'}$$
$$y = p_{y1} x_1 + p_{y2} x_2 + p_{yv} x_v \tag{2'}$$

These residuals are shown on the diagram as follows:

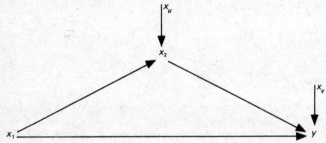

The first residual (x_u) represents factors excluded or ignored in equation (1); the second residual (x_v) represents the factors excluded or ignored in equation (2).

The coefficients $p_{21}, p_{2u}, p_{y1}, p_{y2}$ and p_{yv} can be estimated in either of two ways: (a) by applying ordinary MULTIPLE REGRESSION METHODS to each of the equations (1') and (2'); (b) by the method of decomposition of correlation coefficients, first developed by Sewall Wright (1921 and 1934).

Path analysis is a useful method of measuring the direct influence along each path in the system and thus allocating the variation in a particular variable among the possible causes of that variation. In the example above we would ascertain what proportion of the variation in number of children was explained directly by the age of the women, and what proportion was explained by the effect of age working through its relationship with years of education. (◊◊ DIRECT EFFECT, TOTAL EFFECT, CAUSAL MODEL, PATH MODEL.)

Path coefficient. The coefficients of the variables in PATH MODELS are called path coefficients. They represent the direct effect of one variable on another in the model. Conventionally the variables in the model are assumed to be STANDARDIZED. (◊◊ PATH ANALYSIS.)

Path diagram.   The diagrammatic representation of a PATH MODEL. The relationships between the variables are represented on the diagrams by lines. two kinds of relationships can be represented – directed arrows indicating a causal relationship in the direction in which the arrow is pointing; and curved lines indicating simply an association between two variables. A simple example is given below.

If we are interested in the relationships between race, age, years of education, marital duration and number of children for a sample (or population) of married women, we could represent the model as follows:

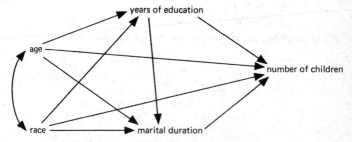

There is a causal sequence implied in this diagram. Age and race are assumed to be prior to (unaffected by) the other variables in the model and thus all the arrows from age and race are pointed towards the other variables. 'Years of education' is next in the sequence and affects marital duration and number of children. Marital duration can be affected by age, race and years of education (so that arrows towards marital duration start from these variables), but can affect number of children (so that there is an arrow from marital duration to number of children). Age and race are not placed in order but they may be related (so they are joined together by a curved line). Numerical values are allocated to the lines in the diagram – these values on the directed arrows are called PATH COEFFICIENTS and are estimates of the strength of the relationship indicated. The manner of estimating the coefficients is described briefly under PATH ANALYSIS. The coefficients of the curved lines are simply the CORRELATION COEFFICIENTS between the variables concerned.

Path diagrams have the advantage of making a model explicit and thus making possible a critical evaluation of the assumptions involved.

Path model.   A CAUSAL MODEL in which the relationships between the variables are represented by magnitudes of effects along paths connecting the variables in the model. The model may be represented either as a PATH DIAGRAM or as a system of equations.

Consider the following example. We are interested in the relationships between the following five variables, using the notation below:

x_1 : age
x_2 : years of education
x_3 : marital duration
x_4 : desired family size
y : number of children

The sample might be selected from the population of women aged between fifteen and forty-five who had at some time been married. The relationships might be specified as follows.

1. The number of years of education a woman completes will be affected by when she was born since there has been a change in education patterns over time; this can be written as

$$x_2 = p_{21} x_1 \tag{1}$$

where p_{21} is the coefficient which expresses the strength of the effect of x_1 on x_2. If the variables are STANDARDIZED the coefficient is called a PATH COEFFICIENT and there will be no constant term.

2. Marital duration (x_3) is affected both by age and by years of education; for example, the more years of education, the later a woman is likely to marry and hence the shorter the marital duration; this can be written as:

$$x_3 = p_{31} x_1 + p_{32} x_2 \tag{2}$$

3. Similarly, desired family size may be affected by age, years of education and marital duration; this can be written as:

$$x_4 = p_{41} x_1 + p_{42} x_2 + p_{43} x_3 \tag{3}$$

4. Finally, number of children (present family size) can be affected by all the other four variables, i.e.

$$y = p_{y1} x_1 + p_{y2} x_2 + p_{y3} x_3 + p_{y4} x_4 \tag{4}$$

The system of equations (1), (2), (3), (4) constitutes a path model. It can be represented diagrammatically as follows:

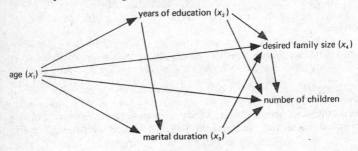

This is an example of a PATH DIAGRAM. The methods of obtaining estimates for the path coefficients are described briefly under PATH ANALYSIS.

Percentiles. The kth percentile is defined as the value of the data such that k per cent of the observations are less than that value and $100 - k$ per cent are greater than that value. For example, the 50th percentile is called the MEDIAN.

Periodicity. The characteristic of happening at regular intervals in time. The term not only relates to cyclical fluctuations and seasonal fluctuations, but sometimes to any wave-like oscillation which has identifiable maxima and minima for each of a series of cycles. Hence although a period should strictly mean a fixed interval of time it is sometimes used to mean any interval of a series in time, of which each of the intervals has a number of given recurring characteristic events or states.

For example, sales may suddenly increase immediately after Easter, yet Easter is not a fixed calendar date, and the cycle between one Easter and the next is therefore somewhat irregular. On the other hand, if the irregularity is such that prediction is impossible, periodicity cannot be said to occur.

Note that in meteorology the term periodicity is confined to cycles which are of regular occurrence. Oscillatory movements which occur at approximately equal-sized but not regular intervals are said to be persistent but not periodic.

Periodogram. There is a long history of attempts to analyse time series in terms of trying to find 'hidden periodicities'. This has been done by a combination of the mathematical methods of Fourier analysis and a statistical approach to the stochastic or random nature of time series. One of the most famous historical attempts was that of Sir Arthur Schuster in 1898: he devised the periodogram to measure the strength of each frequency component in a time series. The periodogram of a time series $x_1, x_2, \ldots x_n$ is given by

$$I(\lambda)^2 = \frac{2}{n} \left[(\sum_{t=1}^{n} x_t \cos \frac{2\pi t}{\lambda})^2 + (\sum_{t=1}^{n} x_t \sin \frac{2\pi t}{\lambda})^2 \right]$$

where λ is the wavelength or period of the cyclic component being studied and $I(\lambda)^2$ measures its amplitude or intensity. By plotting $I(\lambda)^2$ against λ one can display the intensities of any set of wavelengths to show their relative importance.

Permutation. An order in which a given set of objects can be arranged. Thus, for example, the set ABCD may be ordered in the following permutations:

ABCD	BACD	CBDA	DABC
ABDC	BADC	CBAD	DACB
ACBD	BCDA	CABD	DBCA
ACDB	BCAD	CADB	DBAC
ADBC	BDAC	CDBA	DCAB
ADCB	BDCA	CDAB	DCBA

Note that there are twenty-four arrangements, or permutations, of the set ABCD. It can be expressed as $4! = 4 \times 3 \times 2 \times 1 = 24$.

The number of permutations (of r elements) that can be obtained from a set of objects of size n is:

$$\frac{n!}{r!(n-r)!} \times r!$$

that is, the number of combinations of size r multiplied by the number of permutations of each r elements. This simplifies to:

$$\frac{n!}{(n-r)!}$$

To provide two examples:
1. the number of permutations of any two items from a set of four items is:

$$\frac{4!}{2! \times 2!} \times 2! = 12$$

In the set ABCD, these two-item permutations are 12, i.e. AB, AC, AD, BA, BD, BC, DA, DB, DC, CA, CB, CD.
2. The number of permutations of any four items from a group of ten items is:

$$\frac{10!}{6! \times 4!} \times 4! = 210 \times 24$$
$$= 5040$$

or alternatively

$$\frac{10!}{6!} = 10 \times 9 \times 8 \times 7$$
$$= 5040$$

Pictogram. A method of pictorial representation of quantities or magnitudes in a graph, BAR-CHART, band-chart or CARTOGRAM by using a number of miniature drawings of the subject matter, or of related symbols, e.g. £ signs for pounds sterling, barrels for crude oil production, and so on.

For example, the number of employees of Megalith Limited in the years 1973–77 were:

1973	3005
1974	4512
1975	5993
1976	6497
1977	9001

Using a schematic person symbol to indicate 1000 employees, the relevant numbers of employees of Megalith Limited for the years 1973–77 may be illustrated by pictogram in the following way.

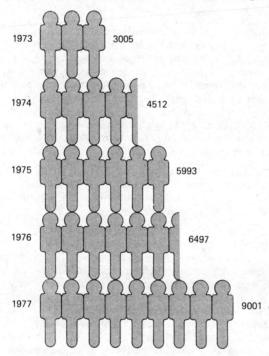

Pilot survey. A small replica of the main survey in which all or most of the survey procedures are pretested. The pilot survey will usually follow a series of PRETESTS in which each of the components is tested separately. The pilot survey then tests the combination of the components under similar conditions but more closely monitored than the main survey conditions can be. No survey should be carried out without a full series of pretests and a pilot survey.

Pivot. The essential element in the inversion process necessary for the solution of a linear program. The pivot is to be found in the column having the highest corresponding coefficient in the objective function and where the quantity Partition column coefficient/Pivot column coefficient is least. Thus in the linear program

Maximize $3x + 4y$
Subject to $x + 2y \leq 100$
$x + y \leq 80$

the highest coefficient in the objective function is 4 (i.e. $4y$) and the pivot row is the second row for 100/2 is less than 80/1. Thus the pivot is 2. (◆ SIMPLEX METHOD.)

Point estimation. The estimation of a parameter by a single value, as distinguished from INTERVAL ESTIMATION, which is the estimation of a parameter, using an interval or range of values.

As the process of estimation is an inexact one, by definition, it is best to regard the difference between point estimation and interval estimation as a technical difference. For example, using point estimation a population mean (μ) is estimated as a single value, with a given standard error ($\sigma/\sqrt{n}$), whereas using interval estimation the upper and lower limits of an interval thought likely to include the parameter being estimated are defined.

Point prevalence rate. The number of current spells related to the number of persons at risk. For example, if on a particular day (date) 8 of the 128 workers in a plant are suffering from influenza the point prevalence rate is $8/128 = 6\frac{1}{4}\%$.

Poisson distribution. A special type of probability distribution which can be applied to a large number of physical and natural phenomena. Suppose there are a large number of occasions during a particular period of time in which some event can occur, and on any one occasion there is a small but identical probability that it will occur, but the average rate of occurrence for identical periods is the same, then the Poisson distribution will give the probabilities that during the period of time the event will occur a specified number of times.

One example of how the distribution can come about is as follows. Consider a sequence of binomial trials in which n becomes large and at the same time p, the probability of success, become smaller in such a way that np remains fixed. The binomial distribution will get closer and closer to the Poisson, and indeed the latter can be thought of as a limiting form of the former. However, many phenomena have been shown to be adequately approximated by the Poisson – the distribution of the number of deaths due to horse-kicks sustained by infantrymen in the Franco–Prussian war; the number of arrivals at a bank teller's desk, the number of phone-calls received by a person during a day, the number of London Transport buses passing a particular road intersection during periods of one hour, and so on.

The probability that a variable with a Poisson distribution takes on a value r is given by

$$\frac{e^{-\lambda} \lambda^r}{r!}$$

where r can take on any positive value, λ is the mean value of r, and e is the exponential (approx 2.718).

Politz–Simmons method. A technique which attempts to eliminate the need for CALL-BACKS in social surveys by using additional information available for those contacted in the first call. The idea is that each respondent who is contacted is asked on how many of, say, five similar occasions he would have been available. His response is then weighted by the inverse of the proportion of times he would have been available in order to compensate for similar people who have not been contacted. For example, if the respondent says that he would have been available on two of five similar occasions, the probability of finding him at home is $(1 + 2)/(1 + 5) = 3/6 = 1/2$. (The extra 1 above and below the line is the call on which contact was made.) Of a hundred similar respondents, the interviewers would expect to contact only fifty. Thus the responses given by those fifty should be given double weight (a weight of 2) in the analysis to take into account the fifty who were not contacted.

The evidence about the effectiveness of this procedure is not favourable but it is probably preferable to having no call-backs and simply ignoring the problem.

Polytomous. Divided into many parts, or categories; as distinguished, for example, from DICHOTOMOUS, which means divided into two parts.

Population. The word population, when used by a statistician, may refer to any specified collection of objects, people, organizations, etc. Some examples of a population are (1) individuals in a country, town or constituency; (2) animals on a farm; (3) commercial firms in a city; (4) record cards in a filing system; (5) children in a school; or, indeed, any other predefined set of elements. The population is the parent group.

Post-enumeration surveys. These are essentially quality checks on the fieldwork or results of a sample survey or a census. Typically, the post-emuneration survey will cover only a part (possibly a small part) of the original population or sample, and the supervision and execution of the fieldwork will be much more rigorously controlled. Such quality checks are now a regular part of censuses, and are becoming more common for all large-scale survey operations.

Posterior probability.   The term means an estimate of probability after a given event. In contrast, prior probability is the estimate of probability before the event. The terms are relative. An estimate which is posterior to a particular trial or experiment would be prior to subsequent trials or experiments.

In essence, the posterior probability is the prior probability modified in the light of observations.

Post-stratification.   A synonym for STRATIFICATION AFTER SELECTION.

Power.   The term has two different meanings:
1. an EXPONENT or LOGARITHM indicating the number of times which a BASE number must be multiplied by itself to obtain a given value, for example the value 128 is 2^7 (i.e. 2 multiplied by itself 7 times), thus the base number 2 is raised to the power of 7; and
2. in statistical testing the POWER of a test is $1 - \beta$, where β is the probability of type II error; that is, its power is the probability that the alternative hypothesis will be rejected when it is false. (♦ BETA-ERROR, TYPE II ERROR, ALTERNATIVE HYPOTHESIS.)

Power function.   The power of a test of the null hypothesis, expressed as a function of a parameter φ which expresses the class of alternatives to it. It may be graphically expressed:

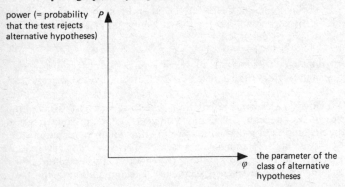

power (= probability that the test rejects alternative hypotheses) P

φ the parameter of the class of alternative hypotheses

Hypothesis tests can be compared by examining the graphs of their power functions.

PPS.   A widely used abbreviation for PROBABILITY PROPORTIONAL TO SIZE.

Precision.   Unlike ACCURACY, which relates to the nearness of a *particular* estimate to a true value of a parameter, precision relates to the nearness of a series of observations to a true value or to each other. Many samples may be taken from the same parent group, and parameters may be estimated from them. The measure of precision is the measure of nearness of these estimates to the true value and their consistency when compared with each other. In short, while accuracy is concerned with single values and is the converse of inaccuracy, precision is concerned with repeated observations of values and is the inverse of DISPERSION.

Pre-coded questions.   Questions to which the response is directly coded into one of a limited number of categories. The question may be a CLOSED QUESTION or may be asked in OPEN-ENDED form with the interviewer carrying out the task of classifying the response into one of the categories. The categories must be mutually exclusive (i.e. should not overlap) and exhaustive (i.e. should cover all possible answers). If a residual category is used, the exact response should be recorded if it falls in this category – the usual form of the residual category is 'Other – please specify'.

Prediction.   The term usually means the estimation of a future value of a variate either (1) by the projection of a trend regression line in time series analysis or (2) by the use of a probabilistic model or (3) both. Unlike estimation, which is mainly used in the literature to mean the determination of an approximate parameter value from a sample, prediction is mainly used in association with linear, multivariate or bivariate regression models, for example $Y = F_t(X)$ where X is a known currently existent independent variable and F_t is a time function. Because it can be used as a means of predicting the future value Y, the REGRESSOR X is sometimes also known as a predictor.

Prediction, error of.   The difference between a prediction and the true value of the identity predicted. Errors may arise because:
1. the model which originally generated the prediction is an incorrect or inaccurate one; or
2. exceptional events have occurred in the time interval which either render the model generally inaccurate or inadequate or are not specifically applicable to the time interval under consideration; or
3. there is a random or stochastic element in the model or process.

Prediction interval.   As stated in the entry PREDICTION (2), estimates of future predicted values may contain a stochastic element and even when they do not do so (◗ PREDICTION, ERROR OF (3)) there is a stochastic or random element in the process itself. Sometimes a prediction is stated as an expected value, accompanied by an expected variance. The alternative method of allowing for a random process is to state the upper and lower limits of a prediction. The interval between the upper and lower values is known as the prediction interval.

Predictive validity. If a scale is constructed as a predictor of behaviour or as an indicator of future performance its success (or validity) can be measured by the degree to which it succeeds in this prediction or indication. This is called predictive validity.

Pre-test. An enquiry or investigation, usually on a small scale, in which some aspect of the methodology to be used in the main study is tested. Typically the questionnaire, the interviewers' method of approach, and other isolated parts of the overall design will be tested separately to begin with; these are the pre-tests. Finally the procedure will be tested as a whole before the main operation is carried out – this is the PILOT SURVEY, a small replica of the main survey.

Primary data. Data obtained direct from source and for the solution of a specific problem in hand. Their essential features are: (1) that they are original, having been obtained by specific investigation rather than from a published, classified or structured source; and (2) that they are obtained for a specific purpose relevant to the enquiry which has been carried out. The principal means of obtaining primary data are by self-administered questionnaires, interviews, observation and experiment.

In historical research, the term 'primary data' is often used to include published material which has not been used by a previous historian, and in accounting, the documents and records from which final accounts are compiled are quasi-primary – that is, primary in nearness to source, but not necessarily obtained solely for the purpose of a specific task (e.g. compiling final accounts) since accounting data are obtained to satisfy a number of differing objectives (e.g. control of debtors).

Primary sampling unit (PSU). The sampling units used in the first stage of multistage sampling are called primary sampling units (PSU). In order to begin a multistage sample selection all that is required is a frame or list of the first-stage units (the PSUs). Subsequently lists may be required of later-stage units but these will be necessary only for the elements within the selected PSUs.

Probability. This term must be considered from two standpoints, its mathematical foundation and the interpretation of probability measurements.

The mathematical foundation of probability in set theory was laid by Kolmogorov and others. Probability is measured in a scale from 0 (= impossible) to 1 (certainty). If two events are equally likely and one or other of them is certain to happen (such as the outcomes 'head' or 'tail' on tossing an unbiased coin), the probability of each of outcome is 0.5. Similarly, if a die is cast, given that it is also unbiased, since it consists of six sides which are equiprobable and it is certain that one side will land uppermost, the probability measurement in respect of each of the outcomes is 1/6. The following scale illustrates these four measurements.

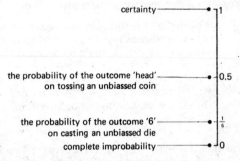

Set theory may be applied to probability measurements. Thus, for example, the probability of either one outcome or another, given that they are independent is the sum of the probabilities of each event. These principles are considered under ADDITION PRINCIPLE (IN PROBABILITY), MULTIPLICATION PRINCIPLE (IN PROBABILITY), etc. At the level of interpretation, however, two extreme positions are adopted, which tend to polarize divergence of views, although there are many positions between the two extremes.

The objective view of probability states, in effect, that mathematical measurement of probability can only be applied to events that can be repeated a large number of times under similar conditions. Thus we may talk objectively about the outcome 'head' being 'probable' to the extent of 0.5 on tossing a coin because it may be demonstrated that in a large 'population' of coin-tossing events, the relative frequency of head outcomes is 0.5. The objective view of probability thus sets aside all unique events, such as the probability that Nero actually set fire to Rome, because the objectivist makes interpretations only from repeated events.

The personalistic view of probability is that the measurement scale that we have discussed applies to measures of 'reasonable' personal belief in a particular proposition. This school of thought would apply the measurement scale to all the kinds of events considered by the objective school, and to many other events. It would, for example, extend the measurement of probability to the belief of 'reasonable' people, given the evidence, whether Nero did or did not set fire to

Rome, and thus apply the 'many event' considerations of the objectivist to one historical (or likely future) event. Accountants and financial advisers take a personalistic view of the probability of future profits. Experts may make estimates of the future profit and prospects of a given firm. A weighted average of the estimates of such experts may then be used as an 'expected value' of the firm's future profits. Similarly, the 'odds' associated with racing winners are to a large extent calculations based on a personalistic view of probability.

Probability density function.   A mathematical expression showing the relative frequency (or probable density) of x as a function of x, i.e. $f(x)$.

For example, the probability density function for the binomial distribution is:

$$f(x) = \binom{n}{x} p^x (1 - p)^{n-x} \quad (x = 0, 1, \ldots n)$$

The values of $f(x)$ for all x, must sum to 1. For a continuous x

$$\int p(x)dx = 1$$

Probability distribution.   The distribution of the probabilities of occurrence of the values of a variate as distinct from a FREQUENCY DISTRIBUTION, that gives the frequencies of the values of the variate.

Example

The probability distribution of outcomes from tossing a coin three times may be represented by the following table.

Value	*Probability*
HHH	0.125
HHT	0.375
HTT	0.375
TTT	0.125
Total	1.000

The mean of the distribution is 0.5, the probability of a head (H) or a tail (T).

Probability proportional to size (PPS).   The probability of selection of a unit (usually a cluster of elements) is proportional to its size (the number of elements it contains). Thus, if we were selecting a sample of manufacturing firms, we might wish to give a probability of selection to the firms proportional to the number of employees. With PPS a firm employing 1000 workers would have twice as high a chance of selection as a firm employing 500 workers, and ten times the chance of selection of a firm employing 100 workers. (◗ SELECTION WITH PROBABILITY PROPORTIONAL TO SIZE.)

Probability sampling.   Any sampling method in which every element in the population has a known non-zero probability of selection. In order to be able to use the sample data to make inferences about the population we must use probability sampling. This is because every element must have some chance of selection (i.e. a non-zero probability of selection) or else the characteristics of those elements cannot be represented by the sample, which may lead to bias in the estimation; and the probability of selection must be known or else we cannot construct sensible estimators based on the sample responses. All the principal sampling methods – simple random sampling, stratified sampling, cluster sampling, etc. – are forms of probability sampling. The term 'random sampling' is sometimes used as an alternative but should be avoided since it can also be taken to mean simple random sampling. Further, it should not be confused with HAPHAZARD SAMPLING.

Probable error.   A measure used in older statistical texts, and still occasionally used in business statistics. It is $(0.6745 \times$ s.e.$)$ where s.e. is standard error. The reason for its use is that when the dispersion of the estimator accords with the NORMAL DISTRIBUTION half of the distribution lies within one probable error on each side of the mean.

Process average proportion defective.   The term is used in quality control to indicate the average proportion of defectives in a batch or sample, and thus the probability that any individual item produced in a controlled process is defective.

Producer's risk.   The risk that a particular batch of satisfactory quality will be rejected by a sampling plan. The risk will depend on the level of quality required and the predetermined criteria of the sampling plan.

Product moment correlation.   ◗ CORRELATION COEFFICIENT, SPEARMAN'S RHO and MULTIPLE CORRELATION.

Profile.   A graphical depiction of a set of values obtained, for example, by assessing scores in the answers to a series of questions.

For example, in psychological testing, a candidate is presented with a mixed series of questions which test aesthetic, academic, social, religious, political or other capabilities or attitudes. Scores are awarded in respect of each of the answers to these questions, and assigned to different letters or codes in each section of the test.

The scores are then added in respect of each of the codes. The resulting scores are plotted on a profile of values and compared with an 'average' profile as in the example below:

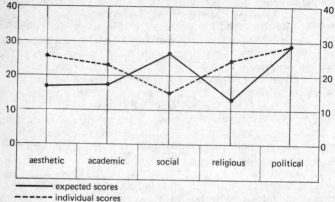

Progression.   An ordered sequence of numbers. For example, the sequence

2, 4, 6, 8, 10, ...

is ordered such that the intervals between each successive pair of numbers are constant (i.e. 2), and is known as an ARITHMETIC PROGRESSION, while the sequence:

2, 4, 8, 16, 32, 64 ...

is ordered such that the ratio between each successive pair of numbers is constant and positive at +2, and is known as a GEOMETRIC PROGRESSION. The essential features of a progression are:
(1) that the numbers are ranked in rising or falling sequence; and
(2) that there are means of predicting successive terms in the sequence.

Progressive average (or mean).   An average (or mean) which is successively modified as additional individual values of a variate are obtained, discovered or become available, such that the progressive mean $\bar{X}_{(n+1)}$ may be valued

$$\bar{X}_{(n+1)} = \frac{n\bar{X}_n + X_{(n+1)}}{(n+1)}$$

Example
The profits of company A for the years 1972 to 1976 in £'000s are:

91, 65, 38, 12, 62

As figures become available the progressive average will be calculated thus:

Year	Available data (£'000s)	Progressive average
1972	91/1	£91 000
1973	(91 + 65)/2	£78 000
1974	(91 + 65 + 38)/3	£64 667
1975	(91 + 65 + 38 + 12)/4	£51 500
1976	(91 + 65 + 38 + 12 + 62)/5	£53 600

The concept of progressive average should not be confused with that of MOVING AVERAGE.

Propensity.   A quantification of a given trend or tendency in terms of proportion, probability or relative frequency. For example, in economic statistics the propensity to save means the proportion of a given income which will be saved, while the propensity to consume means the proportion of a given income which is likely to be consumed. In this case the sum of the two propensities is 1.

The term has a similar meaning in demography, e.g. propensity to marry, propensity to have families, etc., meaning relative frequencies in either case.

Proportion.   The expression of relative magnitude of a subset of a population having a given ATTRIBUTE.

Example
Of the population of a comprehensive school, 600 pupils are boys and 400 are girls. The relevant proportions are:

Boys $\dfrac{600}{1000}$ = 0.6 and

Girls $\dfrac{400}{1000}$ = 0.4

Proportionate stratified sampling.   STRATIFICATION denotes a procedure by which the population is divided up into subdivisions, known as strata, in each of which sampling is carried out independently. Proportionate stratified sampling involves selecting a number of units from each stratum which is proportional to the number of units in the stratum; in other words, we would use the same sampling fraction in each stratum. Consider, for example, the population of a town which is divided into two strata (1) those living in the town centre – say, 100 000 people; and (2) those living in the suburbs – say, 50 000 people. If we wish to select a proportionate stratified sample of 300 people from this population we would use the following procedure:

(a) Overall sampling fraction = 300/150 000 = 1/500

(b) Stratum 1 (town centre): Stratum size = 100 000
Sampling fraction = 1/500

$$\text{Sample size within stratum } 1 = \frac{100\,000}{500}$$

$$= 200$$

(c) Stratum 2 (suburbs): Stratum size = 50 000
Sampling fraction = 1/500

$$\text{Sample size within stratum } 2 = \frac{50\,000}{500}$$

$$= 100$$

(d) Total sample size = 200 + 100 = 300

Proportionate stratified sampling has two major advantages. First, the use of uniform sampling fractions makes the analysis much simpler since the sample is SELF-WEIGHTING. Second, in contrast to disproportionate allocation there is no risk that the precision of the estimators will be lower than that obtained through using simple random sampling.

Proportionate stratification, proportionate stratified sample. With proportionate stratification the same sampling fraction is used in selecting the sample in each stratum. In other words, the number of elements appearing in the sample from each stratum is proportional to the size of the stratum (i.e. the number of population elements in the stratum). Proportionate stratified sampling will be more precise than simple random sampling whenever the mean of the survey variable differs from one stratum to another.

Pseudorandom number. A number taken from a set of numbers which are not truly random, for example, because they are self-repeating or degenerate. (♦ RANDOM NUMBER GENERATION.)

PSU. A widely used abbreviation for PRIMARY SAMPLING UNIT.

Q

Quadratic. Involving second power of unknowns (e.g. x and y), i.e. x^2, y^2. Quadratic equations are those which include second but not higher powers of unknowns (e.g. $x^2 + 4x - 5 = 0$).

Quadratic mean. An infrequently used term which denotes any mean which has been derived by employing the squares of the values for which the mean has been produced. For example, the VARIANCE is a quadratic mean of differences between each of a set of values and their mean, while the STANDARD DEVIATION is also a quadratic mean, the square root of the variance being obtained for a comparative purpose to bring it to the dimension of the differences themselves. The standard deviation is more strictly a quadratic mean than the variance.

Quality control. An application of statistical analysis to the control of quality of a mass-produced article. The reason for using statistical analysis is that the product is manufactured in such large numbers that complete inspection of all articles is impossible. Samples from batches are therefore analysed either: (1) with the object of discovering the number of defective items; or (2) with the object of comparing the mean and standard deviation of product size or component quantity with the desirable size or component quantity, given tolerance limits.

The first kind of test is called attributes sampling. If, for example, the analysis of a sample of 1000 items reveals that 4% are defective, single sampling plan tables will show that there is a 90% chance that the average percentage of defectives in the batch is not larger than 7.99%. The facts of the case will determine whether this is acceptable or not.

The second kind of test is called variables sampling. For example, a rivet must be 2 cm in diameter. A small tolerable discrepancy is allowable at, say, 0.001 cm. The firm wishes to keep well within this limit and samples large numbers of rivets in order to calculate that it can with confidence produce rivets with a mean diameter of 2 cm and a standard deviation of 0.0001 cm. Thus 67% will vary within 0.0001 cm from the mean, 95% will vary within 0.000196 cm from the mean, and 99% will vary within 0.000258 cm from the mean. It is usual to set warning limits at a 95% confidence level and action limits at a 99% confidence level. If a sample of size 100 has a mean value above or below the warning limits at 0.0000196 cm on either side of the mean (2 cm) action will be taken, while if the sample mean is found to be more than 0.0000258 cm above or below the mean of 2 cm the batch is not accepted and each rivet must be examined individually.

Quantiles.   Values at certain positions in a set of numbers arranged in order of magnitude. Common examples are deciles and percentiles.

For example, the third decile is the value below which at least three-tenths of the observations lie or above which at least seven-tenths lie. The twenty-ninth percentile is the value below which at least 29% of observations lie or above which at least 71% lie.

Quantity weights.   When constructing a price index it is normal to form a weighted average of the relative prices of the items covered by the index. These weights are the quantities of the items purchased in some reference period, which could be either the current period for which the index is being calculated or some previous base period. Thus the Laspeyres index can be written as

$$\sum_{i=1}^{n} w_i \frac{p_i^{(1)}}{p_i^{(0)}} \times 100$$

where

$$w_i = \frac{p_i^{(0)} q_i^{(0)}}{\Sigma p_i^{(0)} q_i^{(0)}}$$

i.e. where the weights w_i are the expenditures of the various commodities purchased in the base period (period 0) expressed as a proportion of the sum of all expenditures; $p_i^{(1)}$ and $p_i^{(0)}$ ($i = 1, 2, \ldots n$) are the prices of the commodities in time periods 1 and 0 respectively. An alternative but equivalent way of looking at the same thing is to take a basket of commodities typically purchased in the base period and compare its cost then with its cost when revalued at the prices of the later period. Here the quantities of the base period are used as weights for the prices of the later period, i.e.

$$\frac{\sum\limits_{i=1}^{n} q_i^{(0)} p_i^{(1)}}{\sum\limits_{i=1}^{n} q_i^{(0)} p_i^{(0)}} \times 100$$

A major problem in index number construction is to decide what are appropriate weights to use. As relative prices change, people's purchasing patterns also change so that the cost of one period's purchases in a different period does not properly reflect purely the price movements experienced. If income rises by the same amount as a base weighted (Laspeyres) index it will be possible to switch from purchases of items whose prices have increased relatively more quickly to those whose prices have increased relatively less quickly and by so doing to increase (at least marginally) the total utility achieved from the given income.

Quantum changes. These are the relative changes in quantities purchased between two periods which are used to construct a quantity index number. They play the same role in the quantity index as price changes play in a price index.

Quantum index numbers. A quantum index measures the change between two periods in the average quantity of goods purchased, produced, etc. The typical household, individual, firm, government agency, etc. purchases a wide range of items and we often wish to measure the extent to which purchases have increased in real or quantity terms as distinct from the increase in the value of purchases that is the direct result of price changes. One way to do this would be to deflate the value of purchases by a price index. A better and more direct method is to construct an index of quantities in much the same way as we construct price indices, although, of course, the roles of prices and quantities are reversed. A set of reference prices is chosen – for example, either current period prices or prices for some base period – and quantities of goods purchased in successive periods are then revalued using these reference prices. The sum of the revalued quantities is then divided by its value in the base period and multiplied by 100 to form an index. For example, a Laspeyres quantum index would be

$$Q_1^{(0)} = \frac{\sum_{i=1}^{n} p_i^{(0)} q_i^{(1)}}{\sum_{i=1}^{n} p_i^{(0)} q_i^{(0)}} \times 100$$

where $p_i^{(0)}$ are the reference prices of the base period (period 0), $q_i^{(0)}$ and $q_i^{(1)}$, $i = 1, 2, \ldots n$, the quantities of the various commodities purchased in periods 0 and 1 respectively. $Q_1^{(0)}$ is the index for period 1 based on period 0 = 100.

Quartile deviation. The mean difference between the two quartiles and the median of a set of numbers, or alternatively the INTERQUARTILE RANGE divided by two.

For example, in the following set of numbers:

1, ⑦, 10, ⑬, 15, ㉑, 26

the median is 13 and the lower and upper quartiles are 7 and 21 respectively. Quartile deviation, the mean difference between the two quartiles and the median may either be calculated

$$\left.\begin{array}{l} 13 - 7 = 6\\ 21 - 13 = 8 \end{array}\right\} \text{ mean of 6 and 8 = 7}$$

or alternatively

$$\text{Quartile deviation } = \frac{\text{interquartile range}}{2}$$

$$= \frac{21 - 7}{2}$$

$$= 7$$

The advantages of quartile deviation and interquartile range as measures of dispersion are that, in contrast to range, extreme values do not give a misleading impression of the extent of dispersion (◆ RANGE) and, in contrast to STANDARD DEVIATION, it is not necessary to know all values for the purpose of calculation. It should be added that, although quartiles are easy to define where a set of numbers is large or continuous, the concept of quartile deviation is not particularly meaningful in a small set of numbers such as that given above. The example has been provided simply to show how quartile deviation is calculated. Similar difficulties arise with other quantile measurements.

Quartile measure of skewness. An easily obtainable measure of skewness which employs the fact that in a skewed distribution the median does not lie exactly midway between the quartiles. The extent of skewness can be assessed by doubling the median and deducting this from the sum of the quartiles, $Q_3 + Q_1 - 2M$. To transform this measure into relative terms, it is necessary to obtain the coefficient of skewness, by dividing by the interquartile range. Thus

$$\text{Quartile coefficient of skewness } = \frac{Q_3 + Q_1 - 2M}{(Q_3 - Q_1)}$$

This is the most common quartile measure of skewness. The measure $Q_3 + Q_1 - 2M/\frac{1}{2}(Q_3 - Q_1)$ appears in some texts.

Quartiles. The term is unambiguously defined only for CONTINUOUS DISTRIBUTIONS, for which the quartiles are the three values of the variate which divide the distribution into four equal parts. For discrete or discontinuous distributions, especially those with a small number of elements, some convention must be adopted to make the definition operational. One possible definition is that the upper quartile is the value at or above which one-quarter of the observations lie and at or below which at least three-quarters lie. Similarly, the lower quartile has one-quarter of observations below it and three-quarters above it. The middle quartile is the MEDIAN.

For example, the lower quartile of the numbers 1, 2, 3, 5, 7, 8, 9, 14, 15 is 3.

Quasi-random sampling. The employment of a method of sampling which is not truly random, but where the results will with confidence for practical purposes approximate those which would occur if a truly random sampling process were used. For example, in auditing the use of random sampling is time-consuming, for it involves forward and backward movement through files in order to retrieve all relevant documents. The use of systematic sampling (i.e. the choice of every nth voucher) is often much less costly and produces a similar result (except when the interval n is known to coincide with a period cycle, such that the nth item coincides with a rush period or tea-break). This method of sampling is quasi-random.

Questionnaire. A formulated series of questions. In social surveys a distinction is sometimes made between a questionnaire which is filled in by the respondent himself and an INTERVIEW schedule which is completed by an interviewer.

Questionnaire design. A formulated series of questions used for obtaining information on special points. The function of a questionnaire is measurement, and its specification should state the main variables to be measured. Before constructing the questionnaire the broad structure of the inquiry must be known – whether the population being studied consists of adults or children, for example. A number of decisions must be made before the questions themselves can be written. First, the main and auxiliary methods of data collection must be decided – particularly whether personal interviews or mail questionnaires are to be used. Second, the method of approach to the respondent must be specified. Third, the framework of the questionnaire and the order of topics are important. Fourth, the order of questions within each question sequence and the use of techniques such as funnelling and quintamensional design must be considered. Fifth, the relative advantages of closed and open-ended (pre-coded and free-response) questions must be evaluated.

Each survey presents its own problems and there is no substitute for exhaustive pilot work in designing a questionnaire. All aspects of the questionnaire should be investigated in pre-tests and pilot studies before the questionnaire is used in the main fieldwork of a survey.

Quintamensional plan of question design. A method devised by Gallup of combining open-ended and pre-coded questions in covering a particular issue. The method involves a series of five types of question:
1. Questions which ascertain whether the respondent is aware of the issue and whether he has considered it.

2. Questions designed to obtain the respondent's general feelings on the issue – open-ended questions.

3. Questions on specific topics involved in the issue – usually precoded questions.

4. Questions about the reasons for the respondent's views.

5. Questions to find out the strength of the respondent's views on the issue.

Quota sampling.   A generic term which describes loosely a set of sampling methods used widely by market research organizations in interview surveys. The distinguishing feature of a quota sampling scheme is that each interviewer's workload is based on a fixed quota of individuals who must be representative of the population in respect of some characteristics. The characteristics usually employed are age, sex and socio-economic status. For example, if an interviewer has 100 interviews to complete, the proportion of males in the sample must be equal to the proportion of males in the population; the proportion of respondents over 65 must be the same as the proportion of the population over 65, etc. Given these restrictions, which make up the quota controls, the interviewers then do their best to find people who satisfy the requirements.

Quota sampling has the characteristic, common to all non-probability sampling, that the sampling procedure is ill-defined. There is no way of knowing what chance of selection any individual in the population has had. Within the restrictions imposed by the quotas, the choice of respondent is left entirely to the discretion of the interviewer. This dependence of the achieved sample on the (non-measurable) whim of the interviewer is a source of bias in the selection procedure and thus can give rise to a bias in the estimate derived from the survey. The most important criticism of quota sampling, however, is paradoxical. There is no such thing as a 'representative', 'fair' or 'unbiased' sample. These terms apply only to the procedure by which a sample is drawn and not to the sample itself. Only by a knowledge of how the estimate derived from the sample can vary from one possible outcome of the sampling procedure to another can the procedure be judged. In the case of quota sampling, since the method of selecting the sample is not clearly defined there is no way in which the likely accuracy of the estimate can be discovered. Due to the selection bias, some individuals may have no chance of selection – but there is no way of knowing which individuals these are. Therefore not only is there a non-measurable bias but also the possible fluctuations from one sample to another remain unknown.

Quota sampling is practised mainly because its cost per element is lower than for probability sampling. However, in terms of the precision of the estimates achieved, this reduction in cost may be spurious. The less restrictive the quota controls, and therefore the more scope for selection bias on the part of the interviewer, the cheaper quota sampling becomes. A reduction in monetary cost at the price of increased bias is not necessarily a good bargain!

Despite their faults, quota samples sometimes produce good results. The fundamental problem remains, however, that due to their nature it is not possible to say how well they do perform.

R

R^2. In a bivariate case R^2 is the product of the two REGRESSION COEFFICIENTS b_{xy} and b_{yx} where x and y are the variables. It is also the square of the Pearson product moment correlation coefficient. More generally, it means the proportion of the dependent variable explained by a particular regression equation. If, using a particular multiple regression equation estimates of a dependent variable y are produced and these estimates are then correlated with the observations of y, R^2 is the square of the correlation coefficient between these estimates and observations.

In econometrics the term is understood in this more general way.

Radix. The number upon which a system of numbers is based. For example, the radix of a system of natural logarithms is e (≈ 2.718). The term may in statistical usage mean the base of a formula for the calculation of a table (e.g. 0.3989 in calculating the ordinates of the normal distribution) or it can loosely mean a unit of measurement commonly employed in a table, for example, expenditure on books per 1000 of population in the case of public library statistics.

Raising factor. This term is generally synonymous with one of the meanings of FACTOR, that by which a quantity has to be multiplied to obtain another quantity. It is, however, sometimes used specifically to mean the factor by which a sample value has to be raised to obtain a parameter estimate. Thus, if a simple random sample of size 100 is taken from a population of size 1000 the total earned income, annual expenditure, annual savings, capital formation, etc. of the sample would be multiplied by 10 to estimate the corresponding parameters of the parent group. The raising factor would be 10, and is the inverse of the SAMPLING FRACTION, which in this case is $^1/_{10}$.

Random. A concept very difficult to define uniquely. The term is used in a wide variety of contexts and, if it is to be defined at all, it must be in terms of PROBABILITY. A method of selecting a sample may be said to be random if it gives to each element in the population an equal (or at least, calculable) chance of being selected. However, the term is sometimes used incorrectly to denote 'haphazard'. Formally the use of the term random should mean that the process being described is probabilistic, i.e. depends on the operation of some probability or chance mechanism.

Random error. An error is the deviation between the true value and the observed value or between the true value and its estimate. A random error is an error whose magnitude depends on some probability mechanism, i.e. values occur as though chosen at random from a set of possible values.

Random number generation. The production and use of random number tables has posed several problems. Humanly produced tables are often biased because of the tendency of some people to associate random numbers with odd and prime numbers. Even if random number tables are compiled by drawing numbers from a bag and replacing them, the task of using the method to generate a sufficiently long sequence is tedious. Also, careless use of existing tables often generates bias because of a wrong tendency to commence at the beginning or some other fixed point in the table each time the tables are used.

Computer programs for the generation of random numbers fall into two categories:
1. Those which use an algorithm. (a) The simplest device is the mid-square method, that of squaring the two middle digits of a number to obtain the next number, e.g.

1*12*3, 0*14*4, 0*19*6, 0361 ...
... 16*48*, 40*96*, 0081 ...

The problem with this particular method is its degeneracy, for if it generates 3600, for example, an endless sequence of 3600s is generated. (b) Another possibility is that of repeating sequences. These result from more complicated programs which satisfy the usual tests for randomness and can be used for most statistical purposes, but which tend to be self-repeating after a sufficiently large sequence of numbers has been generated.
2. Radio-active decay pulses. Radio-active decay is considered to be an effectively random process. If a material with a sufficiently long half-life is chosen the strength of the radio-active pulses can be converted into random numbers.

Random numbers. ◗ RANDOM (SAMPLING) NUMBERS.

Random order. The term is commonly used to denote an ordering of a set of objects (a list of names, for example) which results from a process which ensures that every possible ordering has an equal probability of occurrence. However, the term may be used to describe an ordering resulting from a process which ensures that every possible order in the set of objects has a calculable non-zero probability of occurrence. If a set of cards is shuffled correctly and dealt, the order is random; if names are written on slips of paper, shuffled and drawn from a hat, the order of drawing approximates random order.

Random process. A process indicated by a set of variate values distributed either continuously or discontinuously over time, x_t, x_{t+1}, x_{t+2}, ... x_{t+j}, where the change in value from time t to time $t+1$ is independent both of x_t and x_{t+1}. The effect of chance can be simulated in computer models by programming a computer to select from a series of RANDOM NUMBERS or RANDOM STANDARDIZED NORMAL DEVIATES.

Random sample. A sample which has been selected by a method of RANDOM SELECTION. The American term 'probability sample' is probably preferable since the term random may be confused with 'haphazard'. The term random sample is also sometimes used to denote a SIMPLE RANDOM SAMPLE, although this usage is not satisfactory. Random samples are distinguished from QUOTA SAMPLES and any form of MODEL SAMPLING.

Random sampling. A term sometimes used to denote a sampling method in which every element in the population has a known non-zero probability of selection. The term PROBABILITY SAMPLING is more widely used, particularly in the United States, and is preferable since confusion often arises between the term random sampling and SIMPLE RANDOM SAMPLING and since random sampling is sometimes (inappropriately) used to mean HAPHAZARD SAMPLING.

Random (sampling) numbers. Sets of numbers used in selecting (simple) random samples. The sets are generated by a process which involves a chance element and are presented as a series of the digits 0–9 in which each of the digits occurs with approximately equal frequency and, as far as possible, sequences of digits also occur with equal probability. A set of numbers from the 100 000 random numbers published by Kendall and Babington Smith is given below:

```
96 16 76 52 88 95 49 13 21 82 85 84 19 01 03 64 74 91 50 92
01 22 04 38 45 59 91 92 53 20 86 75 18 12 30 15 44 28 22 73
44 11 38 22 82 31 01 46 05 89 36 44 14 07 25 80 80 04 06 77
26 87 15 33 90 55 71 13 93 31 07 30 21 59 71 41 77 03 47 04
49 10 33 76 70 24 35 33 19 69 41 17 60 48 78 72 21 23 44 24
```

If we want to use the random numbers to select a sample of, say, twenty elements from a population of 1000 elements we proceed as follows. The elements in the list must first be labelled from 1 to 1000. We then start at a haphazardly selected point and proceed down each *three-digit* column. Three digits are sufficient here since the number 000 in this case identifies the 1000th element. Here the twenty elements we select are:

```
961 012 441 268 491 676 204 138
715 033 528 384 228 339 767 895
559 231 055 024
```

If the same number occurs more than once in our set of numbers we accept it only if we are sampling with replacement. Otherwise, we reject it and continue until we obtain the desired number of distinct numbers.

If we use the table of random numbers frequently, we may proceed as follows. On the first occasion, start in the top left-hand corner and proceed as described above. When we have finished we mark the point at which we stopped and use the next number as our starting point on the next occasion. Consequently over a period of time we use the full selection of numbers in the table.

Random selection. A method of drawing a sample such that each unit (and each possible sample) has a known non-zero probability of selection. To guarantee randomness some objective mechanism such as a table of RANDOM NUMBERS is necessary. Haphazard selection is not equivalent to random selection. The random or probability element is necessary to guarantee freedom from SELECTION BIAS.

Random standardized normal deviates. A set of random numbers converted into standard score (Z) equivalents. The result is a series of independent drawings from a normal distribution with mean zero and standard deviation 1. In a table of random standardized normal deviates, 68.26% of values will range between -1 and $+1$, approximately 95% between -2 and $+2$, etc. having the relative frequencies of the normal distribution.

Tables of such deviates are often written into computer programs for simulating business conditions. Thus if a company is known to have a likely mean profit of £250 000 per annum with a standard deviation of £50 000, the table of random standardized normal deviates can be converted into profit equivalents £239 000, £281 000, £275 000, ... etc.; by providing equivalent actual random numbers for the particular distribution required, the company's random process of profit over a time span may be simulated.

Random start. In SYSTEMATIC SAMPLING the first sample unit is selected at random from the first set of sampling units – the first SAMPLING INTERVAL. The sample is then said to have a random start.

Random variable. A quantity which varies with a given frequency distribution, so that values occur with specific probabilities. The term random variable is synonymous with variate. (◊ VARIABLE.)

Random walk. A random walk is a stochastic process or a model for a time series that postulates that the value at time $t(x_t)$ is simply the previous value x_{t-1} plus a purely random and unpredictable disturbance e_t. That is

$$x_t = x_{t-1} + e_t \qquad t = 1, 2 \ldots$$

The term is a little misleading as a true 'random walk' would be a two-dimensional affair with steps of random magnitudes being taken in random directions. The model described here, and conventionally referred to as the random walk model, corresponds more closely to the positions at successive moments of time of a drunkard in a narrow alley where he is constrained to move either towards or away from the point from which his position is measured by steps of random length.

The random walk model is often used to describe security prices on a stock market. In such a situation it is to be expected that day-to-day changes are random since, if they were at all predictable, in a perfect market traders would exploit that predictability and so remove it.

Randomization. A process of ordering a set of objects or numbers using a process that ensures that every possible order (or arrangement of items) of the set is equally likely. The simplest illustration of randomization is shuffling a pack of cards. If a pack of cards is well shuffled it is equally likely that any card in the pack will occupy any position (1, 2, 3, . . . 52) in the pack.

Randomness test, runs test. A test for randomness, i.e. to detect a non-random sample. If the sample is one of ordinal, interval or ratio-scale measurements, the median is calculated and observations above or below the median denoted with a + or −. If the sample is one of attributes (e.g. male or female) each observation is denoted as an M or F, or by using other binomial notation. Observations are grouped into runs using the order in which they occur, the number of runs is calculated and also the number of observations in each of the two classes. Tables indicate the probability of obtaining a given number of runs ($P = 5\%$) for particular sizes of the subsamples (x, y; M, F). If there are too many or too few runs the sample is rejected as non-random.

Example
Consider the attribute sequence

   MMMM FF MM FFF M FF

There are seven males, seven females and six runs. Tables show that at $P = 5\%$, where both subsamples have a size of 7, the number of runs may vary between 3 and 13. The sequence can be assumed random. If the number of runs (or clusters) had been less than 3 or more than 13 the order could be assumed non-random and the hypothesis of randomness rejected.

Range. The difference between the largest and smallest of a set of variate values. It is the simplest and most obvious measure of dispersion. For example, if the two distributions

  1, 5, 12, 17, 21, 40

and

  7, 9, 14, 15, 17, 19

are considered, the range of the first series is 39 while that of the second series is only 12, and the first set of values appear to be more widely dispersed than the second series.

  However, although simple, the range may not be an accurate measure of dispersion. The following diagram illustrates two frequency distributions whose measures are both x_{max} and x_{min}, yet that indicated by the curve ———— is much more widely dispersed in terms of variance than that indicated by the curve •—•—•—•

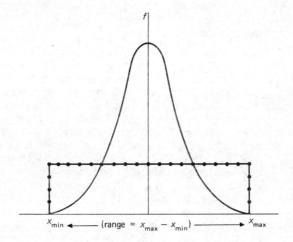

Range may, under normal circumstances, be used for estimating STANDARD DEVIATION (♦ SNEDECOR'S CHECK) or the range of a number of subsamples employed for calculating standard deviation.

Rank (1). The most common and general meaning of the term in statistics is the ordinal position of an observation or value of a variate when such values or observations are arranged in order of a quality or quantity, such as magnitude.

For example, the constituents of the series

> 5, 23, 4, 17, 11, 39, 6, 10

have the ranks

> 2, 7, 1, 6, 5, 8, 3, 4

when arrayed in order of magnitude.

In this general sense, rank is important:

1. in the computation of rank correlation coefficients, such as SPEAR-MAN'S RHO and KENDALL'S TAU;

2. in simple tests of unmatched and matched samples, such as WIL-COXON'S SIGNED RANKS and SUM OF RANKS TEST and the KRUSKAL AND WALLIS TEST;

3. in concepts associated with comparative magnitude in series of numbers, such as median, quartiles, deciles, centiles, etc.

Rank (2).   The term is sometimes also used in matrix theory to denote the greatest number of linearly independent ROWS or columns which can be found in a matrix.

Rank correlation.   A loose term which covers a series of procedures for measuring how closely different sets of rankings of a collection of objects or individuals agree or disagree.

Example

One judge in a beauty contest may rank six contestants Miss A, Miss B, Miss C, Miss D, Miss E, Miss F, 2, 3, 1, 5, 6, 4, where the smaller the rank the more the contestant is preferred. Another judge may rank them 1, 2, 3, 6, 5, 4. Rank correlation measures the extent of agreement or disagreement between the judges. If the judges were in perfect agreement then any measure of rank correlation would take a value of 1. If the judges were in perfect disagreement, as they would be if the first judge ranked the contestants 1, 2, 3, 4, 5, 6 and the second ranks were 6, 5, 4, 3, 2, 1, then they would take conventionally the value -1. Between these limits there are degrees of agreement and disagreement. A value of the rank correlation coefficient of around zero would indicate that the criteria being applied by each judge in making his ranking bear no relation to and were independent of those of the other. A measure of rank correlation is called a rank correlation coefficient. Those in most common use are KENDALL'S TAU and SPEARMAN'S RHO.

Ratio.   An expression of relative magnitude between two or more values in a series or between elements of the series and their total. For example, the ratio between 63 and 36 is $7 : 4 (= 1.75 : 1)$.

Ratios are used in accounting as heuristic guidelines for the purpose of testing liquidity (e.g. Liquid assets/Current liabilities) or profitability (e.g. Yearly profit/Capital invested).

Ratio estimator.   An estimator which takes the form of a ratio between two random variables. Such ratios are often used in sample surveys. The value obtained by using a ratio estimator is known as a ratio estimate.

Examples
The number of books borrowed from libraries in an area may be estimated from the ratio between total borrowings in a sample set of areas (10 000 000 per year) and the total population of the sample set (750 000). If an area consists of 30 000 people the ratio estimate of books borrowed per year is

$$30\ 000 \times \frac{10\ 000\ 000}{750\ 000} = 400\ 000$$

Ratio estimators are frequently used where ancillary information (last year's data, for example) is available for the whole population and observations on a sample only are available for this year. In this case the estimator takes the form

$$\hat{Y}_R = \frac{y}{x} \times X$$

where y and x are this year's and last year's sample totals respectively, X is last year's population total and $\hat{Y}_R$ is the ratio estimator of this year's total.

Ratio scale (1). A SCALE OF MEASUREMENT which has the properties of an INTERVAL SCALE together with an absolute zero or origin. Thus there exists a point on the scale which corresponds to the absence or zero value of the characteristic. Height, weight and time are examples of ratio scales. With such scales we can order the scale scores, we can compare distances between them and we can compare the absolute magnitudes of the scores. For example, on a scale of length we can compare the scale scores 24 inches and 12 inches in three ways: (1) 24 inches is longer than 12 inches (the ordinal property); (2) the distance between 24 inches and 12 inches is the same as the distance between 36 inches and 24 inches (the interval property); (3) 24 inches is twice as long as 12 inches (the ratio property). Most physical measurement is on a ratio scale but in attitude measurement we must normally be satisfied either with ordinal or INTERVAL SCALES. All statistical techniques can be used on data measured on a ratio scale.

Ratio scale (2). A graphical scale in which equal absolute variations correspond to equal proportional variations in the data. The most commonly used is in the LOGARITHMIC or SEMI-LOGARITHMIC CHART. Such a scale is used when relative rather than absolute changes are considered to be the important consideration.

Ratio test (or variance ratio test). ◗ F-TEST.

Raw data. The data from a survey or experiment in their original form, i.e. before being modified or analysed in any way. Raw data are, however, often assumed to be edited and coded. ◗ CODE.

Recall errors/memory errors. In answering a factual question the respondent usually relies to some degree on memory to provide an accurate answer. There are two basic types of recall errors. The first is RECALL LOSS, or memory failure, when the respondent does not report an activity because he has forgotten about it. The second is TELESCOP-ING, the process whereby the respondent includes in the reference period some event which occurred outside it.

Recall loss. In questions dealing with past events, the respondent may simply have forgotten some events and therefore fail to report them. This is recall loss. The two most important factors in determining when such recall loss will occur are (1) the length of time since the event occurred and (2) the saliency, or importance, of the event to the respondent. In requesting information about the past, therefore, great care should be taken that information is not sought which it would be unreasonable to expect the respondent to remember. The interview situation, especially if the respondent does not have access to documents or records, may not be the most suitable method of data collection for this kind of information.

Recode, recoding. Substituting new CODES for the original coding of the data. In coding the responses to a survey, the original codes may either be unsatisfactory or too detailed for some of the analysis. In such cases, the data may be recoded by using a computer program. For some types of analysis the data must be in a particular form and some standard programs are available to modify (recode) the data.

Record check. A method used occasionally to check on the accuracy of the information collected in a survey. There may be some external source (record) for some of the information obtained. A check on age reporting may be carried out by examining the register of births, for example. However, there are two serious drawbacks to these procedures in general. First, it will not usually be possible to obtain records for all the respondents, and even when available they may be expensive to obtain. Second, if there is a conflict between the two sources of information it may not be possible to ascertain which of them is correct. In short, the external source of information must be current, comparable and reliable, and for practical purposes, accessible. Such information is not usually available.

Rectangular distribution. A frequency distribution in which the frequencies of all possible values of a variate are identical and therefore fx_i is a constant where x_i is any possible value of variate x. It is called a rectangular distribution because it is depicted graphically by means of a single rectangle. It must not be confused with RECTANGULAR HYPER-

BOLA. The use of a single rectangle is best employed to depict a continuous rectangular distribution. In the case illustrated below there is a certainty that a value lies between 15 and 25, and there are equal probabilities (each of 0.1) that it lies between 15 and 16, 16 and 17, 17 and 18, etc.

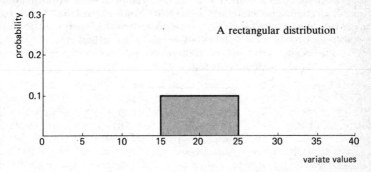

A rectangular distribution

Rectangular hyperbola. A locus of points so drawn that the area of any rectangle confined by any pair of ORDINATE and ABSCISSA lines (i.e. lines drawn parallel with the ordinate and abscissa axes) and the appropriate parallel axes is a CONSTANT.

Rectangular hyperbolae are asymptotic – they converge with their adjacent axes in the limit, i.e. at infinity. Indifference curves, used in economic analysis, generally take the form of rectangular hyperbolae.

Example

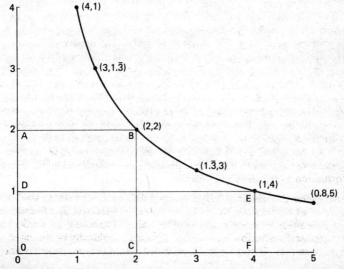

Note that the product of the co-ordinates at all the points is 4, (i.e. 4 × 1, 3 × 1.$\bar{3}$, 2 × 2, 1.$\bar{3}$ × 3, 1 × 4, 0.8 × 5) and that therefore the area confined by parallel ordinate and abscissa lines and axes at any of the co-ordinate points (e.g. OABC, ODEF) is a constant (4).

Refusal rate.   In a SAMPLE SURVEY we attempt to obtain information from all eligible selected sample elements. Some of these may refuse to provide the information (i.e. refuse to respond). The proportion of those eligible elements who refuse is called the refusal rate.

For example, if a sample of 1000 people were selected, the outcome of the fieldwork might be as follows:

Ineligible	100
Refusals	60
Not at home	50
Not reached	40
Respondents	750
Total sample	1000

Of the 900 eligible sample members, 60 refused. This gives a refusal rate of

$$\frac{60}{900} \times 100\% = 6.7\%$$

Refusals.   One of the categories of NON-RESPONSE, those who, although successfully contacted, refuse to give the information requested. Refusals will arise due to a number of reasons, some of which are transient – indisposition, annoyance at being approached at an inconvenient time – and some of which are more stable – some people may object in principle to being asked questions by a stranger and may consider an interview to be an invasion of their privacy. The latter attitude is not, however, very common and in many cases an alternative approach may produce a successful interview. Flexibility is therefore important in approaching a potential respondent and every effort should be made to motivate the respondent to cooperate although a 'hard core' of refusals will almost always remain.

It is also possible that a respondent who cooperates with the survey may refuse to answer particular questions. This should be coded separately in the data when it occurs.

Regressand.   A term sometimes used instead of dependent variable in regression analysis. It may be regarded as the antithesis of REGRESSOR, as dependent variable is the antithesis of independent variable. The terms regressand and regressor are to be preferred since the terms dependent and independent carry with them an implication of causation, which is not a necessary criterion in regression analysis.

Regression analysis. In any process or system in which quantities change, it is of interest to examine the effects that some variables have (or seem to have) on others. The actual relationship between the variables may be extremely complicated and we generally try to approximate the relationship by some simple (or fairly simple) mathematical function or equation. The equation may be substantively meaningful or may be used simply as a means of predicting some of the variables from knowledge of other variables.

Two main types of variables can be distinguished: independent (or predictor or REGRESSOR) variables and dependent variables. The independent variables are those which can either be fixed in advance (e.g. experimental controls) or can be observed but not controlled (e.g. temperature or rainfall). We observe the effect of changes in the independent variables on the dependent variables. In general we are interested in the way in which the independent variables affect the dependent variables. We assume a simple type of relationship, usually linear, and estimate the unknown parameters with the help of available data.

The LEAST SQUARES METHOD is frequently used to estimate relationships in the data. The overall method of analysis is called regression analysis and the equations (lines) estimated are called REGRESSION CURVES or REGRESSION LINES.

Regression coefficient. The coefficient of any regressor (or independent variate) in a regression equation. Thus, if the regression equation is simply $Y = a + bX + \varepsilon$, the regression coefficient is b.

Regression constant. This term is sometimes used to indicate the constant (usually described by either the letter a or the symbol β_0) in a regression equation. In cost analysis it is analogous to 'fixed cost'. For example, if the price charged by a motor rental company is £10 per day irrespective of usage and 5p per mile, the price charged to any day-customer (Y) is $£10 + 0.05\,X$ where X is the number of miles travelled. In this case the fixed cost (or regression constant) is £10.

Regression curve. A diagrammatic depiction of a two-variate regression equation, using a locus of points (either linear or non-linear) which are co-ordinates for the dependent variable Y and the regressor X. Y is usually measured on the ordinate axis and X on the abscissa. Because of the scattered nature of the co-ordinate points the curve is usually a 'curve of best fit', the unexplained or stochastic differences being indicated by the term ε.

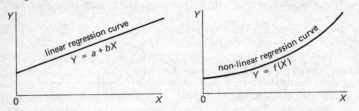

A regression curve can strictly define the relationship between only two variates, but if three-dimensional, isometric or computer models are used the term may be loosely employed to mean the depiction of an equation where there are more than two variables (or variates).

Regression estimate. An estimate of the value of a dependent variable Y for given values of the independent variable(s), $X_1, X_2, \ldots X_j$, using a regression equation. ($\blacklozenge$ REGRESSION ANALYSIS.)

Regression line. A synonym for REGRESSION CURVE in most general usage where the regression curve happens to be linear. ($\blacklozenge$ LINEAR REGRESSION, REGRESSION ANALYSIS.)

Regressor. An independent variable in a regression equation or relationship. Regressors are usually expressed as $X_1, \ldots X_j$ in contradistinction to the independent variable Y.

Rejection. In statistics the term rejection is applied to the rejection either of a HYPOTHESIS, or in the case of QUALITY CONTROL (q.v.) of a set of items of a given quality. Although the two contexts in which the word is most frequently used seem to be unrelated, in both cases rejection assumes that a sample (in hypothesis testing) or a batch (in quality control) has characteristics or values which render it improbable, at a given *significance level*, that it conforms with the requirements.

Relationship. An expression of dependence between two or more variables in contrast with, for example, a correlation coefficient, which expresses the measure of interdependence. Contrast, for example, the coefficient $r_{xy} = 0.89$, which expresses the measure (or extent) to which x and y are associated, with the relationship $y = 252 + 7.6\,x$, which indicates how the two variables are associated.

Relative deviate. The extent of deviation between two absolute values usually expressed as a fraction of the mean of all observed values.

Example
A particular value of x is 60 and the mean of all observed values is 50.

The value could be expressed as an ABSOLUTE DEVIATE $(60 - 50 = 10)$ or as a relative deviate $(10/50 = 0.2)$.

Relative frequency. The frequency of occurrences (or observations) of a particular group (such as the interval class of a variate value) expressed as a fraction of all occurrences (or observations) in all groups.

For example, the age distribution of a youth club is:

Age interval (yrs)	Frequency	Relative frequency	
Under 14	1	1/64	(= 0.015625)
14 and under 16	6	6/64	(= 0.093750)
16 and under 18	15	15/64	(= 0.234375)
18 and under 20	20	20/64	(= 0.312500)
20 and under 22	15	15/64	(= 0.234375)
22 and under 24	6	6/64	(= 0.093750)
Over 24	1	1/64	(= 0.015625)
Total	64	1	(= 1.000000)

The probability of a person selected at random from the club being in a particular age group may be equated with the relative frequency in that group. For example, the probability that any person selected randomly from the total population of the club is aged 18 but under 20 is 0.3125.

Relative likelihood. Maximum likelihood estimates give the most plausible value of a parameter for a specified population distribution. However, other values might be marginally less plausible and in some conditions may be entertained as possible values. A relative likelihood function is used to compare other values with the maximum likelihood value. If four heads were observed in tossing a coin ten times in identical conditions, the maximum likelihood estimate of p the long run proportion of heads in the population of infinite tosses is 0.4. The relative likelihood of any other value of p is $p^4(1-p)^6/(0.4)^4(0.6)^6$. The function ranges in value from 0 for completely implausible values of p to 1 when p is the maximum likelihood estimate. The relative likelihood of $p = 0.5$, for example, is 0.817. This latter value of p although not the maximum likelihood estimate, is still then very plausible.

Relative variance, relvariance. The variance of a distribution divided by the square of the arithmetic mean. Thus

$$\text{relvariance } (x) = \frac{\text{variance } (x)}{\bar{X}^2}$$

The relvariance is the square of the coefficient of variation. Its purpose is to provide a comparative measure of the dispersion of frequency distributions.

Repetition. Denotes the carrying out of a survey or experiment at different places or time periods. This is different from REPLICATION, in which the time and place are kept constant as far as possible.

Replacement, sampling with; sampling without. A sample is said to be selected with replacement if an element which has been selected in the sample is permitted a further time (times) in the sample. In the case of sampling without replacement no element is permitted to be selected (appear in the sample) more than once. For example, in selecting Premium Bond prize winners, winning bonds are selected without replacement. Sampling with replacement would imply that once a winning bond has been selected it would be replaced in the drum before the next winner was selected, thus giving it another chance of being selected. In sampling without replacement, a winning bond is left out of the drum once it has been selected and consequently could not win twice on a particular occasion.

Replicated sampling. Equivalent to INTERPENETRATING SAMPLING, but the term is most often applied to such sampling when carried out for the purpose of providing a simple method of calculating SAMPLING ERRORS. In essence, instead of selecting a single sample a number of subsamples are selected, each of which is a self-contained sample of the population. The variability of the overall estimates can be estimated by comparing the values obtained for each of the replicates (the subsamples).

Having a small number of replicates gives some advantages: (1) the selection procedures may be simplified; (2) more STRATIFICATION can be used; (3) the visual (graphical) presentation of the results is easier; (4) the computation of sampling errors is also easier; (5) it is easier to build studies of non-sampling errors (e.g. INTERVIEWER VARIANCE) into the design.

However, having many replicates also gives advantages: (1) the estimates of sampling error are more precise; (2) the sample design is less susceptible to peculiarities in the population structure.

The choice of the number of replicates depends on the purposes of the study and on the number of PRIMARY SAMPLING UNITS to be selected. When this number is small the method of PAIRED SELECTIONS – two replicates – is often the most appropriate.

In STRATIFIED SAMPLING replicated sampling implies that the same number of replicates is selected from within each stratum.

Replication. The execution of a survey or experiment more than once. The purpose is generally to increase precision and to improve the estimation of SAMPLING ERROR. Replication is different from REPETITION in that in replication, the repetition or duplication is carried out in the same place and at the same time period as far as possible. For example, in REPLICATED SAMPLING the subsamples are all dealt with at the same time. However, replication is often used loosely to mean simply repetition.

Representative.   A term used to describe a SAMPLE which is similar to the population from which it was drawn in terms of some important or relevant characteristics. It can also be used to describe a method of selection which gives an equal chance of selection to all possible samples. This ambiguity makes it an unsatisfactory term.

Representative sample.   In a broad sense, a SAMPLE which adequately represents the population, i.e. it mirrors some important characteristics of the population. Sometimes the term is used to describe a sample which is selected by a sampling process which gives each element or each sample the same probability of selection (◗ EPSEM SAMPLING). It is also used to describe a sample which is similar to the population in some important respects, no matter how the sample is selected (◗ BALANCED SAMPLES). This confusion in usage suggests that, as a technical term, it is better avoided.

Residual.   The difference between an observed and an estimated value. Using an existing model for the purpose of estimation (e.g. $y = f(x)$), the residual term is an expression of the extent of difference between estimated and observed values, and thus the distribution of the residuals is a useful measure for assessing the risk of using a particular equation (e.g. a regression equation) for predictive purposes.

Respondent.   An individual who agrees to participate in a survey and provides information (not necessarily accurate) in response to questions.

Response.   The reaction of an individual unit to a stimulus. In the context of sample surveys of human populations it is the reaction obtained to a request for information (i.e. being asked a question).

Response bias.   The discrepancy between the 'true value' that is being measured and the expected survey result is called the response bias. The measurement of response bias requires information external to the sample itself – in effect it requires that we know the 'true values'. Thus the response bias can be estimated only by intensive measurement or by comparing the survey results with other data which make it possible to evaluate the magnitude of the bias. (◗ RESPONSE ERRORS, INTERVIEWER BIAS.)

Response errors.   Errors which arise during the process of obtaining a response from an individual in surveys of human populations. Such errors may arise in a variety of different ways.

If the respondent is asked how much money he spent on repairs and maintenance in the home in the previous three months, he may not be able to remember accurately which expenditures occurred during the three-month period and may include in his answer expenditures which took place more than three months before, or exclude expenditures which did fall in the period. Such memory or RECALL ERRORS – in this case 'telescoping' – are one type of response error.

In answering questions on attitudes, opinions or beliefs the respondent may be influenced by the apparent social class or attitudes of the interviewer and may therefore give an inaccurate answer. This is a type of INTERVIEWER EFFECT and is another form of response error.

When a question concerns a sensitive or embarrassing topic – drinking habits or sexual behaviour, for example – the respondent may deliberately give an inaccurate answer. Some topics are particularly prone to response errors of this kind. There is sometimes a tendency on the part of respondents to give a socially acceptable answer or to give the answer the interviewer appears to expect.

If the respondent gives the true answer but the interviewer, through oversight, either fails to record it or else records it incorrectly, a response error occurs.

Response errors can be divided into two classes – VARIABLE and constant, or SYSTEMATIC, errors. Faulty recording of answers, for example, is likely to cause positive and negative errors about equally often and the errors will therefore generally cancel out. The same may be true to some extent of memory errors although it is likely that overall there will be some under-reporting of events. On the other hand, there are some errors which are likely to have a systematic effect and which will not cancel out. People in general tend, for example, to understate their consumption of alcohol and tobacco and thus an estimate of alcohol consumption based on their responses will be a serious underestimate. The interviewers themselves may also introduce systematic errors. A classic example occurred in the case of the opinion polls used to predict the result of the 1948 presidential election in the United States. The polls predicted that Dewey would defeat Truman easily in the election. The result was the reverse. In an investigation of the failure of the polls one of the contributory factors identified was the fact that the interviewers, being mainly middle-class college-educated married women, caused some of Truman's supporters – from politeness or a wish to conform – to state a preference for Dewey, the Republican candidate.

The INDIVIDUAL RESPONSE ERROR is the inaccuracy in the response of a particular individual. Response errors include all such errors, however caused. (◆◆ INTERVIEWER BIAS, INTERVIEWER VARIANCE, RESPONSE BIAS, RESPONSE VARIANCE, INDIVIDUAL TRUE VALUE.)

Response rate. In a SAMPLE SURVEY an attempt is made to obtain information (the response) from all eligible selected sample elements. Those eligible elements who provide the information are the respondents. The proportion of all those eligible elements in the sample who do respond is called the response rate. For example, if a sample of 1000 people were selected, the outcome of the fieldwork might be as follows:

Ineligible	100
Refusals	60
Not at home	50
Not reached	40
Respondents	750
Total sample	1000

Of the 900 eligible sample members, 750 responded. This gives a response rate of 750/900 × 10% = 83.3%.

Response variance. Apart from RESPONSE BIAS, the remaining part of the total response error can be thought of as due to variations in responses, or response variance. Response variance is a measure of the variability of those contributions to response error that tend to cancel out with sufficiently large samples. (◆ RESPONSE ERRORS, INTERVIEWER VARIANCE.)

Revaluation at constant prices. When a variable is measured in terms of monetary units it is difficult to distinguish real or volume changes from changes due to price movements. A common practice in constructing national accounts and economic data generally is to deflate values by dividing by a price index. The quantities so obtained are said to be at constant prices of the base year (i.e. base of the price index), and comparisons can be made in real terms. The technique is similar to constructing a Laspeyre quantum index in which base period prices are used to revalue a group of components, but unlike an index number it does not involve expressing successive years' volumes as a percentage of the base year.

Robust. A test procedure is robust if it is insensitive to departure from one of the general assumptions on which it is based. Often a particular significance test depends on assumptions about the parent distribution from which a sample is assumed to have been taken. If one of the assumptions is not correct, but the test procedure nonetheless yields correct results, it is said to be robust (i.e. healthy and insensitive) with respect to that assumption.

Rounding error. The error which results from the process of rounding, i.e. by omitting some end digits and replacing by zeros if required.

Some principles in respect of rounding error are:

1. The maximum rounding error of a sum equals the sum of the maximum rounding errors of its components:

$$(8000 \pm 500) + (900 \pm 50) = 8900 \pm 550$$

2. The maximum rounding error of a difference equals the sum of the maximum rounding errors of its components:

$$(8000 \pm 500) - (900 \pm 50) = 7100 \pm 550$$

(◗◗ ACCURACY, APPROXIMATION.)

Row. A horizontal array of elements in a matrix or table. Thus in the matrix

$$\begin{pmatrix} 2 & 3 & 4 \\ 7 & 12 & 1 \end{pmatrix}$$

the arrays (2, 3, 4) and (7, 12, 1) are rows and (2, 7), (3, 12) and (4, 1) are columns.

It is a normal convention to designate rows before columns. For example, the above matrix is defined as a 2×3 matrix, because it has two rows and three columns. Similarly the letters i and j are used as subscript indices to designate particular rows and columns in the matrix, either indicating a particular element a_{ij}, (e.g. $a_{11} = 2$, $a_{12} = 3$, $a_{21} = 7$, etc.) or a particular row or column (e.g. row $a_{11} \ldots a_{1j}$, or column $a_{11} \ldots a_{i1}$).

Row percentage. The percentage of any particular row in a given cell. For example, if the row total is 400 and a particular cell value is 64, its row percentage is 16%.

Row total. The total of all elements in a row (i.e. $\sum_{i=1}^{k} a_i$ where a_1, $a_2, \ldots a_k$ are the elements of the row).

S

Sample Any subgroup of the population can be called a sample. Samples arise in practice for a number of reasons. In opinion polling, it is too expensive to obtain the views of all the individuals in the population, consequently only a subgroup or sample of individuals is interviewed. In industrial testing of, say, explosives, if all the explosives manufactured were to be tested none would be left. Thus it is necessary to test only a subset or sample.

Sample census. A slightly unsatisfactory term. The word census has two meanings: (1) a complete enumeration of a population and (2) an enumeration in which only basic or straightforward data are collected. It is only when 'census' is used in the latter sense that the term *sample census* has any meaning. In this case it means a partial enumeration (or SAMPLE SURVEY) in which only basic data are collected. The 1966 Census of Population, for example, in Great Britain was based on a 10 per cent sample, and could therefore be described as a sample census.

Sample design. Sample design has two components – the selection process and the estimation process. The selection process consists of the rules, procedures, and operations by which some members of the population are included in the sample. The estimation process consists of the rules and operations for computing sample estimates of the population parameters or population values.

A good sample design will provide a balance between several criteria. First, the sample design should be such that it satisfies the survey objectives. Second, the design should be practicable, i.e. it must be possible to carry out the operations required to select the sample as defined; there is no virtue in an elegant theoretical sample design which is impossible to operationalize. Third, the sample design should be economical, or efficient, i.e. the desired precision and/or accuracy should be obtained at minimum cost; conversely, for a fixed cost, the information collected should be maximized. Fourth, the design should be measurable, i.e. it should be possible to obtain estimates of the SAMPLING VARIANCE of the estimates from the sample itself. These criteria are frequently in conflict and there is no unique best answer to any practical sampling problem.

Sample selection. The process by which we designate the population elements which are to be included in the sample. In probability sampling, an objective mechanism or device such as a table of random numbers is used to obtain the sample. In purposive or model sampling, the interviewer or field worker may have considerable discretion in designating the selected units.

Sample size. The number of units included in a sample. It generally refers to the number of elements in the sample. In MULTISTAGE SAMPLING, for example, the sample size is the number of final stage units in the sample.

Sample space. A sample of n observations can be represented as a point or vector in n–dimensional space. Such a representation is called a sample point. The set of all possible sample points constitutes the sample space.

Generally the term is used in statistics to denote the set of all possible outcomes to a sample selection procedure, or to an experiment. If a coin is tossed twice, for example, there are four possible outcomes:

A head followed by a tail	HT
A head followed by a head	HH
A tail followed by a head	TH
A tail followed by a tail	TT

The set of four possible outcomes in this case constitutes the sample space.

A further example is provided by selecting two elements by simple random sampling from a population of four elements. There are $^4C_2 = 6$ possible samples. If the population elements are denoted by A, B, C and D the possible samples are

$$AB, AC, AD, BC, BD, CD$$

In this case, the order of appearance of the elements is immaterial and the sample space is given by the set of samples:

$$\left\{ AB, AC, AD, BC, BD, CD \right\}$$

Sample statistic. A term which is synonymous with STATISTIC; any measure calculated on the basis of the observations in the sample.

Sample survey. A study of a sample or part of a population (usually used in the context of human populations or aggregates of social or economic institutions) – carried out in order to estimate properties of the population using the sample observations. Opinion polls and some market research studies are carried out using sample surveys. The term should be confined to situations where the sample is selected by probability methods, although it is sometimes used to describe surveys where QUOTA SAMPLING or some other form of MODEL SAMPLING is used.

Sample unit.   Denotes any one of the units constituting a specific sample. It is sometimes used synonymously with SAMPLING UNIT but should be confined to the sense above.

Sampling.   Sampling is the selection of a part of a population for some investigative purpose. A desire for knowledge about a universe or population so large or dispersed that it is generally impracticable or impossible to enumerate or investigate each element in it has led to the use of sampling in almost all branches of human endeavour. Tea tasting, blood testing, quality control, the development of friendship or acquaintance, conversation and even thought are all examples of applied sampling.

The advantages of sampling over complete enumeration are that sampling reduces costs and saves time and labour. Over and above these advantages, a sample may provide more accurate information than a complete enumeration. The quality of the information collected may be much higher than that which could be collected for the whole population since resources are spread over a much smaller operation.

A discussion of the principal types of sampling is given in the entries SAMPLE DESIGN, PROBABILITY SAMPLING, QUOTA SAMPLING, PURPOSIVE SAMPLING, QUALITY CONTROL.

Sampling bias(es).   BIASES which arise in some sense from the sampling process or from the fact that the estimates are based on a sample of the observations. FRAME biases may well be included here, since they usually arise due to failure to adjust estimates for unequal selection probabilities for the elements – arising perhaps from duplication of elements in the list. With PROBABILITY SAMPLING the biases can often be removed by the use of modified estimation procedures.

Sampling biases can also arise through the use of biased but consistent estimators – estimators which are biased for a sample but for which the bias decreases with sample size and disappears for a 100% sample. The RATIO ESTIMATOR is an example. Such biases may be tolerated if the estimator has desirable attributes which are considered to compensate for the bias – high precision, for example.

Sampling distribution.   The array of possible values for a specific sample design, of a sample statistic or estimator, each with its associated probability of occurrence, constitutes the sampling distribution of the statistic or estimator.

Example

A population consists of six elements, i.e. $N = 6$, A, B, C, D, E and F with the following values:

A	B	C	D	E	F
3	5	7	7	9	11

We wish to estimate the mean of the population using as our estimator the mean of a simple random sample of size $n = 2$. There are $^6C_2 = 15$ equiprobable samples ($\blacklozenge$ COMBINATORIAL).

These samples, together with their means, are given below.

Sample	AB	AC	AD	AE	AF	BC	BD	BE	BF	CD	CE	CF	DE	DF	EF
Observations	3,5	3,7	3,7	3,9	3,11	5,7	5,7	5,9	5,11	7,7	7,9	7,11	7,9	7,11	9,11
Sample mean	4	5	5	6	7	6	6	7	8	7	8	9	8	9	10

The possible values of the sample mean are 4, 5, 6, 7, 8, 9, 10. Their frequency and probability distribution is given by:

Value	Frequency	Probability
4	1	1/15
5	2	2/15
6	3	3/15
7	3	3/15
8	3	3/15
9	2	2/15
10	1	1/15
All	15	1

The sampling distribution of the sample mean in this case is that given by columns 1 and 3 of the table above. It is the distribution of the sample means of all possible samples of size 2 from the population of size 6.

The artificial example above illustrates the general principle of the sampling distribution. A realistic example, such as the sampling distribution of the sample mean for all samples of size 2000 from a population of size 3 800 000, would fill a book if all possible samples were written down. Fortunately, however, this would not be necessary since we are generally interested in only two characteristics or parameters of the distribution:

1. The mean of the sampling distribution which is called the *expected value* (q.v.) of the estimator. Thus, for the sample mean, this would be written as

$$E(\bar{y}) = \sum_{c} P_c \bar{y}_c$$

where $\bar{y}$ denotes the estimator, $\bar{y}_c$ denotes the estimate for sample c, and P_c is the probability of occurrence of the sample c. In the example above we have:

$$E(\bar{y}) = 4.(1/15) + 5.(2/15) + 6.(3/15) + 7.(3/15) + 8.(3/15) + 9.(2/15) + 10.(1/15)$$
$$= 4/15 + 10/15 + 1\tfrac{3}{15} + 1\tfrac{6}{15} + 1\tfrac{9}{15} + 1\tfrac{3}{15} + 10/15$$
$$= 105/15 = 7$$

Note that $\bar{Y}$ (or μ), the population mean, is

$$\frac{3 + 5 + 7 + 7 + 9 + 11}{6} = \frac{42}{6}$$
$$= 7$$

$E(\bar{y}) = \bar{Y}$ indicates that $\bar{y}$ is an UNBIASED ESTIMATOR of $\bar{Y}$.

2. The variance of the sampling distribution, which is called the SAMPLING VARIANCE of the estimator.

Sampling error.   Since only a part of the population – i.e. the sample – is observed, we can expect a difference between the estimate obtained from the sample and the population value being estimated. This difference constitutes the sampling error of the estimate. If we consider all possible samples of a given size from the population, the estimates derived from these samples make up the sampling distribution of the sample statistic. The variance of the sampling distribution (or sampling variance) is sometimes described as the sampling error of the distribution. (◆ STRATIFIED RANDOM SAMPLING, CLUSTER SAMPLING.)

Sampling fraction.   The proportion of the total number of units in the population which are included in the sample. For SIMPLE RANDOM SAMPLING this definition is straightforward. In the case of STRATIFIED SAMPLING the sampling fraction should be calculated and presented separately for each stratum, although of course uniform sampling fractions may be used. For MULTISTAGE SAMPLING there is a sampling fraction corresponding to each stage of selection. The application of the term in general to complex sampling schemes is not relevant. Sometimes the definition above is applied to all forms of sampling but this is not satisfactory in the case of selection with arbitrary probabilities. Giving a single sampling fraction for the whole sample should be confined to cases where the sample is SELF-WEIGHTING.

Sampling frame. This consists of available descriptions of the population in the form of lists, maps, directories, etc., from which SAMPLING UNITS may be constructed and a set of units selected. The nature of available frames is an important consideration in determining the sample design. A perfect sampling frame is one in which each element (or unit) appears separately on the list, appears once and only once, and no elements appear on the list which do not belong to the population being studied. Very few frames are perfect, and they may be inaccurate, incomplete, inadequately described, out of date, or some elements may appear more than once on the list.

In MULTISTAGE SAMPLING there is a frame corresponding to each stage of sampling. One of the advantages of multistage sampling is that frames for the later stages need only be constructed for the units selected in the earlier stages. This is particularly true for AREA SAMPLING where the detailed maps and listings necessary for the later stages need only be constructed for the areas selected at the first stage of sampling.

Sampling frames include procedures that account for all the sampling units without actually listing them. Thus a sampling frame for industrial workers could consist of a list of firms, without any initial listing of the workers themselves.

Sampling interval. In SYSTEMATIC RANDOM SAMPLING the fixed interval that is used to generate the sample units subsequent to the first is called the sampling interval. The sampling interval is computed as

$$I = \frac{\text{total number of units in the population}}{\text{desired sample size}}$$

The first unit to be selected is obtained by choosing a unit at random between 1 and I. Once the first unit has been designated, every Ith unit thereafter is also designated in the sample.

If we want a sample of fifty elements from a population of 15 000, the sampling interval will be

$$I = \frac{15\ 000}{50}$$
$$= 300$$

We first choose a random number between 1 and 300. Suppose 77 was chosen. The seventy-seventh case on the list would be the first sample selection. Then merely adding 300 identifies the next sample case, and we continue adding 300 until the elements on the list are exhausted. The first nine selections in this case would be

　　77, 377, 677, 977, 1277, 1577, 1877, 2177, 2477 ...

Sampling plan. The steps to be taken in selecting a sample from a given SAMPLING FRAME. Thus the sampling plan is a part of the overall

Scalar 233

SAMPLE DESIGN, which includes both the selection process and the estimation process.

Sampling unit. One of the units into which the population is divided for the purposes of sampling. The sampling units contain the elements, but the units are considered to be indivisible while the selection is being made. The division of the population into units may be made on any basis but natural groupings of elements are usually used – households, areas, institutions or production batches, for example. In ELEMENT SAMPLING the sampling unit contains only one element; in CLUSTER SAMPLING the sampling unit will typically contain several elements. Within the same survey many different sampling units may be used, as in MULTISTAGE SAMPLING, where the units are different at different stages of selection (see primary sampling unit, secondary sampling unit). The term SAMPLE UNIT is sometimes used synonymously, but has also a different meaning and is thus less desirable.

Sampling variance. The VARIANCE of the SAMPLING DISTRIBUTION of an estimator. This is a measure of dispersion of the estimator and thus indicates how precise the estimator is. The variance is the mean of the squared deviations of the elements from the mean of the distribution. As an example, we calculate the variance of the SAMPLING DISTRIBUTION described in the entry of that name.

Calculation of the sampling variance of the distribution given under sampling distribution

Value	Deviation from mean (=7)	Squared deviation from mean	Frequency	Frequency × squared deviation
4	−3	9	1	9
5	−2	4	2	8
6	−1	1	3	3
7	0	0	3	0
8	1	1	3	3
9	2	4	2	8
10	3	9	1	9
			15	40

$$\text{Sampling variance (= variance of sampling distribution)} = \frac{40}{15}$$

$$= 2.6\overline{6}$$

(◗ SAMPLING ERRORS.)

Scalar. A number as distinct from a VECTOR or MATRIX.

Scale. A standard of measurement consisting of a graduated arrangement of units. The term is used in a more specific sense in graphical representation to mean a set of gradations at measured distances on a line (such as an axis) where interval distances are proportionate to the values which they are intended to represent. Scales of measurement may be associative, ordinal or cardinal, and cardinal scales include ARITHMETIC SCALES and RATIO SCALES. (◆◆ SCALES OF MEASUREMENT and MEASURE, MEASUREMENT.)

Scale point. A particular point on a scale.

Scales of measurement. When a numerical mark or code is allocated to an individual for a particular characteristic, it is necessary to be able to define the meaning of this code in relation to the scores of other individuals in the population. The most elementary type of scale is called the NOMINAL SCALE, where the score simply indicates the group to which the individual belongs. For example, in considering temperature we could allocate scale marks 1, 2, to the categories 'hot', 'cold' in any way we wish. The numbers in this case merely identify the categories and have no other significance. Thus any set of numbers could be used.

The next type of scale is the ORDINAL SCALE, where the scores indicate the rank or order of the elements or groups. In a more sophisticated description of temperature we might have four categories – 'Very hot', 'Hot', 'Cold', 'Very cold' – and allocate the scores 1, 2, 3 and 4 to these. In this case the categories can be *ordered*, each category is hotter than the categories on its right and colder than those on its left. Any set of numbers which preserve the order relationship between the categories can be used to describe the scale points.

The next stage in temperature measurement might be to have a scale where the differences between categories can also be compared. The Fahrenheit and Celsius (centigrade) temperature scales are of this kind and are called INTERVAL SCALES. Thus we can say that the difference in temperature between 75°F and 65°F is the same as the difference between 54°F and 44°F in terms of the scale (10°F). The same property holds for the Celsius scale. However, it is not true that 60°F is twice as hot as 30°F. To be able to make a statement of this kind we have to have a RATIO SCALE. In thermodynamics, temperature measured on the absolute scale has this property – there is a fixed origin at −273°C and relationships between the temperature, pressure and volume of gases can be defined. The most important attribute of the ratio scale is that there is a fixed origin. Length and weight are simple examples. In both cases, we can define the zero point, and comparisons of the form 'twice as' or 'half as' can be used for scale positions. Thus 4 ft is twice as long as 2 ft; 10 tons is twice as heavy as 5 tons. If all scale positions are multiplied or divided by a constant, the scale retains all its characteristics. We could measure weight in pounds rather than tons and length in inches rather than feet – 22 400 lb is twice as heavy as 11 200 lb and 48 inches is twice as long as 24 inches.

The highest form of scale is the ABSOLUTE SCALE. Counting, viewed as measurement, is an example. Here the numbers have an absolute meaning – no other numbers can be substituted for them without destroying the scale.

Physical measurement is usually on a ratio, or at least an interval, scale. The placing of horses in a race, or candidates in an examination, is on an ordinal scale. The numbers given to competititors in a dancing competition represent a nominal scale. In attitude measurement we try to attain the highest scale of measurement we can, although usually we must be satisfied with an ordinal scale – one individual is more authoritarian or less authoritarian than another.

Scatter diagram. A diagram showing a cluster of points representing observations of corresponding values of two variates, such as x and y, such that the two corresponding observed values are co-ordinates of each point on the corresponding axes.

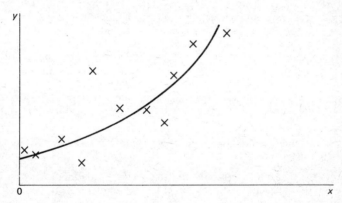

Scatter diagrams are useful aids in estimating the type of regression function which best expresses the relationship between two variables. For example, in the above case the regression function is probably curved.

Scedasticity. A term, not commonly used, which denotes dispersion. It arises now only in the terms HOMOSCEDASTICITY (equal variances) and HETEROSCEDASTICITY (unequal variances).

Schedule. A set of questions in a sample survey, arranged so as to obtain information from an individual (or set of individuals) on a given subject. The term is usually employed in respect of interviews where it need not necessarily be a formal questionnaire, but a kind of check-list of the questions which an interviewer will ask. A structured formal set of questions, designed to be completed by either the interviewer or respondent (i.e. in an interview or by post), is also known as a QUEST-IONNAIRE (◗ INTERVIEW.)

Score. A rating of an individual's performance in quantitative terms or an assessment of some characteristic of an individual on a quantitative scale.

Seasonal adjustment. Many economic quantities vary in a regular cyclical manner with a fixed period. In such cases it is often required to isolate the purely seasonal variation from the trend and other components of the variable. This is achieved by estimating seasonal factors and then subtracting (or dividing by) the factor appropriate to the season in question. Many methods are available and these differ in the structure that they assume for the time series being adjusted. (◗◗ SEASONAL FACTORS, SEASONAL VARIATION.)

Seasonal factors. In a time series that exhibits seasonal variation, the seasonal component that gives rise to this variation can be described in terms of seasonal factors. There are several ways that these factors can be thought to operate and can be built into a model for the series. The simplest model is an additive one where each season is above or below the level due to other influences (trend, cyclical, random) by an amount that is constant for that season each year. A multiplicative model would assume that each season is a constant percentage higher or lower than the level that would hold in the absence of seasonality. For example, consumers' expenditure might be 5% higher than normal in the fourth quarter of each year because of buying for Christmas. Thus we might have a model

Consumer spending = trend × cyclical factor × seasonal factor × random term

where the seasonal factor in the fourth quarter is 1.050. For quarters I, II and III the factors might be 0.944, 1.000 and 1.006 respectively. These factors could be estimated for a time series by first smoothing the series to obtain the trend, then taking the ratios of the series to its trend value, and finally averaging these ratios for each quarter separately.

Seasonal variation. Periodic behaviour in a time series whose period is exactly a year is referred to as seasonal variation. It is caused by climatic variation over the year or by some other phenomenon linked to the annual calendar. If we postulate a model for the series incorporating seasonal factors, then these can usually be estimated to allow us to quantify that part of the total variation in the series that is due to seasonal influences and to adjust the series seasonally if we wish to remove those influences. Many published economic series are given in both adjusted and unadjusted form. Changes in the adjusted figures can be interpreted as changes in the underlying behaviour of the economy and are probably more useful for many purposes.

Secondary data. Data which have been removed at least one stage from source. They are structured or processed in some way. Published documents, such as government reports, sets of statistics issued by the Central Statistics Office, ten-yearly census of population statistics, local government statistics, the published accounts of companies, theses of researchers and other reports come into this category.

The main advantage of secondary data is the comparatively low cost of access, but because the objectives in obtaining the material are different they are less likely to be relevant to particular research than primary data would be.

Selection bias. There is selection bias in a procedure for sample selection which systematically discriminates against some part of the population and gives either no chance of selection or an unknown chance of selection to that section of the population. QUOTA SAMPLING, for instance, gives considerable discretion to the interviewer in selecting respondents, and the way in which interviewers make use of this flexibility may lead to a systematic underrepresentation of some types of respondents. The unknown nature and extent of the selection bias is what causes the problem. In disproportionate stratified sampling, on the other hand, we give different probabilities of selection to the elements within the different strata but since we know what these probabilities are we can adjust for them in our estimates. The probabilities of selection need not of course be equal for all elements or samples as long as they are known. The advantage of probability sampling is that it guarantees freedom from biases in the selection procedure. ($\blacklozenge$ PROBABILITY SAMPLING, MODEL SAMPLING.)

Selection with arbitrary probabilities. In order to develop a general theory of SURVEY SAMPLING it is necessary to be able to cope with the problems of estimation when the elements (or sampling units) within the whole population or within a particular stratum are selected with unequal probabilities, which are determined in advance of the selection.

The general theory is based on selection with arbitrary probabilities – i.e. the case where the probability of selection of a sampling unit can take any value. The two most common forms of selection with unequal probabilities in practice are SELECTION WITH PROBABILITY PROPORTIONAL TO SIZE and SELECTION WITH PROBABILITY PROPORTIONAL TO ESTIMATED SIZE. The statistical theory of sampling can, however, deal with any set of probabilities of selection for the units. If the units are selected one by one, as is often the case, the set of probabilities associated with each drawing may be difficult to compute.

Selection with equal probabilities. A SAMPLE DESIGN in which all the elements in the population or stratum under consideration are allocated the same probability (chance) of being selected into the sample. This is in contrast to selection with arbitrary (unequal) probabilities. Selection of elements with equal probabilities leads to an EPSEM SAMPLE. The principal advantage of an epsem sample is that the estimator of the population mean is self-weighting (i.e. it can be calculated as the simple unweighted arithmetic mean of the sample elements). Basically the term selection with equal probabilities refers to the selection of a single element from a set of elements in such a way that the probability of selection of each of the elements in the set is the same. The generalization of the term to the selection or more than one unit is not uniquely defined. For example, in DISPROPORTIONATE STRATIFIED SAMPLING, although within each stratum selection is carried out with equal probabilities, the probabilities of selection differ between strata. This might be considered to be a case of selection with unequal probabilities but the usage is confused on this point.

Selection with probability proportional to a measure of size, selection with probability proportional to estimated size. When exact values are not available for the sizes of the sampling units (the number of elements they contain), it is impossible to use SELECTION WITH PROBABILITY PROPORTIONAL TO SIZE. However, estimates of the sizes of the sampling units may be available and we can select the sampling units with probabilities proportional to these estimates of size. To the extent that these estimates are inaccurate there will be some loss to the design – either some of the control of the sample size must be sacrificed or else the sample will not be EPSEM.

Selection with probability proportional to size (selection with p.p.s.). A sampling procedure in which the probability of selection of a sampling unit (usually in this case a CLUSTER of elements) is proportional to its size (p.p.s. – in this case the number of elements it contains). (◆ PROBABILITY PROPORTIONAL TO SIZE.)

The main purpose of selecting samples in this way is that in TWO-STAGE SAMPLING selection with p.p.s. at the first stage combined with selecting an equal number of elements from each selected PRIMARY SAMPLING UNIT leads to an equal probability sample of elements – an EPSEM SAMPLE.

The probability of selection of the ath sampling unit at the first stage is proportional to $S_a/\underset{a}{\Sigma}S_a$, where S_a is the size of the ath unit and $\underset{a}{\Sigma}S_a$ is the sum of the sizes of all the sampling units, i.e. the size of the population.

If b elements are selected from within each selected first-stage unit the probability of selection of an element unit α is b/S_a. Therefore, the overall probability of selection of each element in the population is the product of the probabilities at each of the two stages (♦ MULTIPLICA-TION RULE OF PROBABILITY.)

$$\text{Probability of selection of each element} \propto \frac{S_a}{\sum\limits_{a} S_a} \times \frac{b}{S_a} = \frac{b}{\sum\limits_{a} S_a}$$

and is the same for every element in the population   This procedure also gives the researcher control over the final sample size while maintaining equal probabilities of selection. The size of each sampling unit (the number of elements in each cluster) must be known in advance. When exact information is not available, we can use selection with PROBABILITY PROPORTIONAL TO ESTIMATED SIZE (q.v.).

Self-representing p.s.u.   Some primary sampling units (p.s.u.s) are so large that they are automatically selected into the sample. This will occur in sampling with PROBABILITY PROPORTIONAL TO SIZE if a primary sampling unit is larger that the inverse of the SAMPLING FRAC-TION. The sampling procedure with regard to such units consists only of applying the appropriate sampling fraction within them. Some examples of self-representing p.s.u.s might be large cities in a national sample of adults; or large firms in a sample of manufacturing industry.

Self-weighting estimator.   An ESTIMATOR which, for the sample design being used, gives equal weights to the sample observations.

Self-weighting sample.   A sample can be described as self-weighting if the simple mean of the sample observations is an unbiased estimator of the population mean. In general, if the RAISING FACTORS of the sample units are all equal, the sample is self-weighting with respect to the particular estimator under consideration. Sample designs in which the elements in the population are given equal probabilities of selection (EPSEM SAMPLES) normally lead to self-weighting samples and greatly simplify the work involved in tabulation and estimation. In TWO-STAGE SAMPLING, the number or proportion of second-stage units to be selected can be determined in such a way as to make the sample self-weighting.

Semi-averages, method of.   A rapid, but approximate, method of estimating regression lines. The data are divided into two halves, using the median value of x as a partition point. The means of the variables (x, y) for each of the two halves are then calculated, and their co-ordinates plotted on a graph. The two points are then joined and the line linking both points is taken as an estimate of the regression line $(y = a + bx)$

Example	*'Lower half'*		*'Upper half'*	
	x	y	x	y
	3	8	10	18
	5	10	11	20
	7	14	15	25
	8	15	16	27
Semi-averages	5.75	11.75	13	22.5

The least-squares regression equation estimated using all eight points (co-ordinates) is $y = 3.38 + 1.47x$. The estimate obtained using the method of semi-averages, which may be estimated by plotting and joining the co-ordinates (5.75; 11.75) and (13; 22.5) is $y = 3.25 + 1.48x$.

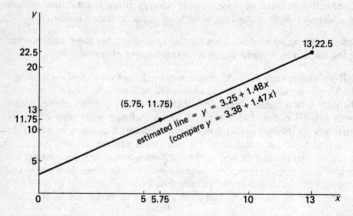

In this case the two estimated equations are in close agreement. However, in general the short-cut method may not be quite as effective as it appears in the example.

Semi-logarithmic chart. A chart in which one scale – usually the vertical (ordinate) – is proportional to the logarithm of the variable being represented and the other scale (abscissa) is proportional to the arithmetic value of the second variable. It is often used to depict a series that is growing at a constant rate through time, which appears as a straight line whose slope depends on the rate of growth.

Sequential analysis.   Analysis of data obtained by sequential sampling or CONTINUOUS SAMPLING methods. Each stage of analysis determines whether sampling is to be continued or not.

Serial correlation.   The correlation between members of a time series and members of a corresponding time series lagged by a fixed interval of time.

If t is the time interval, serial correlation would involve correlating x_1 with y_{t+1}, x_2 with y_{t+2}, x_3 with y_{t+3}, etc., the essential feature of serial correlation being that x_i and y_i are not the same set of variables. In contrast, AUTOCORRELATION involves correlation between x_1 and x_{t+1}, x_2 and x_{t+2}, etc., the values being associated with the same variable, so that whereas serial correlation is a lagged correlation of one variable with another, autocorrelation is a lagged correlation of one variable with itself.

Some authors define serial correlation to include autocorrelation – though not, of course, vice versa.

Serial design.   A form of experimental design involving the introduction of small changes over a period of time; (1) to separate the effects of particular variables and (2) to render the system amenable to continuous review so that the experiment may be extended over a long period.

Sigma (Σ).   When used in mathematical formulae the capital Greek letter S (Σ and pronounced 'Sigma') means 'the sum of'. It is a shorthand method of writing expressions which would otherwise be cumbersome.

For example:

ΣX means: 'Add together all values of the variable X'

ΣX^2 means: 'Square all values of variable X, and then add the resultant quantities together'

whereas

$(\Sigma X)^2$ means simply: 'Square the quantity ΣX'

(as above).

Hence, if the values of variable X are 1, 5, 8 and 9,

$$\Sigma X = 1 + 5 + 8 + 9 = 23$$
$$\Sigma X^2 = 1 + 25 + 64 + 81 = 171$$

while $(\Sigma X)^2$ is simply $23^2 = 529$.

Sign test.   ◗ WILCOXON'S SIGN TEST.

Signed ranks test.   ◗ WILCOXON'S SIGNED RANKS TEST.

Significance.   The term is used mainly in the context of hypothesis testing. It is said to exist when a value lies outside an acceptable interval (known as a confidence interval) and when, consequently, there are good reasons for rejecting a NULL HYPOTHESIS. For example, a null hypothesis that there is no difference between a sample mean and that of a parent group could be upheld at a given level of confidence if the difference between the two lies within a given confidence interval. If the difference between the two means lies outside that interval it is said to be significant. (◊ SIGNIFICANCE TESTS.)

Significance level.   In hypothesis testing it is usual to obtain from a given set of sample data a test statistic calculated for the purpose of the test. This test statistic can only be used if its distribution under the NULL HYPOTHESIS is known.

If the test statistic falls in a range of values, known as the critical region, which, in total, have a small probability of occurrence under the null hypothesis, that hypothesis will be rejected. This small probability is called the significance level.

The most commonly used values are 0.05 and 0.01 (i.e. 5% and 1%), although any other level may be chosen. Thus the significance level is the probability of rejecting the null hypothesis when it is true and is also called the probability of TYPE I ERROR. (◊◊ SIGNIFICANCE TESTS).

Significance tests.   Typically a researcher can study only a small fraction of the population in which he is interested – in statistical terminology, he studies a sample from the population of interest. For example, in medical research a drug company may wish to find out how effective one sleeping tablet is compared to another. Instead of giving each drug to every adult in the population (the population of interest) and observing the effects, it is necessary to confine the procedure to a small number of people (the sample). In the same way, when an organization wishes to predict the result of an election it would be ideal to interview everyone on the Electoral Register and ask their views. Again, however, for reasons of practicality it is necessary to confine the interviewing to a relatively small number of people (the sample).

Any quantity calculated on the basis of the elements of the sample is called a statistic. The difficulty arises in deciding how much faith should be placed in a sample statistic. Consider the examples above. If the two drugs are A and B, say, and drug A induces eight hours sleep for the twenty people to whom it is given and drug B induces nine hours sleep for the twenty people to whom it is given, can we say that drug B will, in general (i.e. for the whole population of interest) induce more sleep than drug A? Or is the difference simply due to random variations in the population? In other words, if we had chosen different people in the sample, would we have got the same answer? Similarly, in the case of an opinion poll, typically about 1500 individuals would be interviewed. Assume 52% of those say they intend to vote for the Labour

Party. Can we say that Labour will get more than half the votes or may this result be due to the particular sample chosen, i.e. could the sample plausibly have occurred if the NULL HYPOTHESIS were true? This is the type of question for which statistical hypothesis testing is designed. Having set up a HYPOTHESIS – e.g. that there is no difference between the effects of drugs A and B – we use the data provided by the sample to calculate how likely it is that the hypothesis is false.

Similarity, coefficient of.   Any measure used to describe the closeness or similarity of elements, groups or variables. The term is used particularly in CLUSTER ANALYSIS. The most commonly used coefficient of similarity is the product-moment CORRELATION COEFFICIENT. However, rank correlation coefficients (e.g. SPEARMAN'S RHO (ρ) and KENDALL'S TAU (τ)) are also frequently used. The coefficients will normally take values between -1 and $+1$. A measure of difference between elements, groups or variables may be called a COEFFICIENT OF DISSIMILARITY.

Similarity matrix (matrix of similarities).   A MATRIX which presents the COEFFICIENTS OF SIMILARITY for each pair of elements in the data set. For example, if we have three elements and the coefficient between elements 1 and 2 is 0.8, the coefficient between 1 and 3 is 0.6 and the coefficient between 2 and 3 is 0.5, then the matrix below would represent the situation

	1	2	3
1	1.0	0.8	0.6
2	0.8	1.0	0.5
3	0.6	0.5	1.0

The coefficient of similarity between an element and itself is equal to 1. The matrix is symmetrical in the sense that the coefficient in the second row of the first column is equal to the coefficient in the first row of the second column. Thus no information would be lost if only the lower triangle in the matrix were presented as

	1	2	3
1	1.0		
2	0.8	1.0	
3	0.6	0.5	1.0

This is the form in which the information is usually given.

Simple random sampling.   A sampling scheme in which each possible combination of n different elements has an equal probability of selection, which implies also that each element has an equal probability of selection. If the population consisted of the four elements $\{1, 2, 3, 4\}$

and we wished to select a sample of two different elements from this population, the set of possible samples would be

(i)	{1, 2}	(iv)	{2, 3}
(ii)	{1, 3}	(v)	{2, 4}
(iii)	{1, 4}	(vi)	{3, 4}

In simple random sampling each of these six samples would be given an equal chance of selection – for example, by throwing a fair die. It is obvious that this also gives each element the same chance of selection; each of the elements appears in three of the six samples and therefore has a probability of ½ of being selected. It would be noted that in this definition the sampling is assumed to be done without replacement, i.e. no element can appear more than once in a sample. Simple random sampling with replacement is referred to as UNRESTRICTED RANDOM SAMPLING.

Simple random sampling is the most basic selection process. All other sampling methods can be seen as departures from it and must be justified on the grounds of economy, precision or practicality. Thus STRATIFICATION is used to increase the precision of the estimators or for administrative convenience. CLUSTER SAMPLING is generally introduced to decrease the cost of the fieldwork or to make sampling possible where it would otherwise be impractical. Unequal probabilities of selection may be necessary to cope with problems in the SAMPLING FRAME or for the purposes of OPTIMAL ALLOCATION. Finally, more than one phase of sampling (MULTIPHASE SAMPLING) may be desirable in some circumstances.

Simple sampling.   Another term for UNRESTRICTED RANDOM SAMPLING. This term, no longer widely used, implies a sampling scheme in which at each successive drawing each element in the population is given an equal chance of selection. Thus sampling is with replacement and an element may appear more than once in the sample. (◆◆ PROBABILITY SAMPLING.)

Simplex method.   A method of solving a mathematical program by the process of iterative application of rules associated with matrix inversion until an optimal solution has been reached.

Example
Consider the LINEAR PROGRAM:

Maximize $Z = 3x + 4y$
Subject to $\quad x + 2y \leqslant 100$
$\qquad\qquad\quad x + y \leqslant 80$
$\qquad\qquad\quad x, y \geqslant 0$

The meaning of each of the functions is explained in the entry on linear programs.

The two inequalities can be expressed as equalities by the addition of slack variables (s and t), such that the program reads

$$- 3x - 4y \qquad\quad = \quad 0$$
$$x + 2y + s \qquad = 100$$
$$x + y \qquad + t = \quad 80$$

This may be written in a partitioned matrix:

x	y	s	t	Z (quantity)
-3	-4	0	0	0
1	②	1	0	100
1	1	0	1	80

A pivot is chosen where the quantity

$$\frac{Z}{\text{More profitable variable}}$$

is least.

And as y is the more profitable variable the pivot is 2.

The pivot is reduced to 1 and all elements of the pivot row are divided by the pivot element. These are then multiplied by the other elements of the pivot column and subtracted from the other rows:

x	t	s	y	Z
-1	0	2	0	200
½	1	½	0	50
⑫	0	$-½$	1	30

The same procedure is again followed and the base figure of ½ in column x is the pivot, such that $x = 60$ and $y = 20$.

	s	t	x	y	Z
	0	0	3	2	260
$y =$	0	1	0	-1	20
$x =$	1	0	1	2	60

The solution is to produce $60x$ and $20y$, making a profit of 260.

Simulation model. Certain types of DYNAMIC MODEL problems cannot be solved by ordinary analytical methods because conditions change rapidly or because of the large number of variables. In such circumstances, it is possible that the problem can be approached by assuming the likely effects of the conditions and producing a computer model which will show the effects of changing one or more of the variables. For example, computer models of business firms can be constructed to simulate the likely effects of trade recessions, seasonal fluctuations, theft, weather conditions, alteration of interest rates, etc.

Such models may either be (1) diagnostic or (2) predictive. The output from diagnostic models is compared with actual business results and the causes of differences examined and identified. Predictive models are used to estimate future results, and are used in budgetary control. (◆◆ MONTE CARLO METHOD.)

Simultaneous. This term means 'at the same time', and is used commonly in two contexts:
1. Simultaneous equations and simultaneous equation systems, where the relationships between a number of variates are expressed by a group of simultaneous equations:

$$a_1x_1 + a_2x_2 + \varepsilon_1 = T_1$$
$$b_2x_2 + \varepsilon_2 = T_2$$

... etc.
(◆ ECONOMETRICS.)
2. Simultaneous estimation, where several parameters are estimated at the same time from the same statistical data.

Single linkage. A form of CLUSTER ANALYSIS. This principle on which it operates is as follows. First, the two elements with the highest similarity in the set are linked together. The next largest value in the matrix identifies the next pair to be linked. The principle used in the process is that an element is linked to a cluster if it is sufficiently similar to any of the elements in the cluster. This is why the method is sometimes referred to as nearest neighbour analysis – an element is included in the cluster to which its nearest neighbour belongs. The process continues until the required number of clusters is formed.

Consider the examples of a similarity matrix below. There are seven elements and we wish to form two clusters.

	1	2	3	4	5	6	7
1	1.00	0.90	0.20	0.15	0.80	0.30	0.20
2	0.90	1.00	...	...	0.75	...	...
3	0.20	0.4	1.00	0.50	...	...	0.70
4	0.15	...	0.50	1.00	...	...	0.85
5	0.80	0.75	...	...	1.00	...	...
6	0.30	...	...	...	...	1.00	0.83
7	0.20	...	...	...	...	...	1.00

The first two elements to be linked are elements 1 and 2; their similarity coefficient is 0.90, the largest in the matrix. The next linkage is between elements 4 and 7 – the coefficient being 0.85. The process continues until two clusters are formed.

Single linkage can lead to a situation where widely dissimilar elements are found in the same cluster. With the points in diagram 2, page 65, for instance, the two clusters formed are those indicated by the two lines. This is known as a chaining effect. If we wish to avoid this problem we can use COMPLETE LINKAGE ANALYSIS

Single sampling.   In evaluating the quality of a product or material by inspecting a part rather the whole, single sampling refers to the method in which the decision whether to accept or reject the batch is taken after the inspection of a single sample. This may be contrasted with sequential sampling, where successive samples (elements or groups of elements) are drawn until sufficient evidence is available to make a decision.

Size.   ♦SAMPLE SIZE.

Skew.   Asymmetric; the term is generally employed for frequency distributions.

Skew(ed) distribution.   A frequency distribution is skew when the first moment (or centre of gravity) does not coincide with other measures of central tendency, because of the asymmetric shape of the distribution. The diagram below illustrates a case of positive skewness.

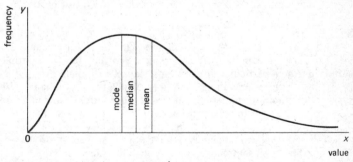

Skewness.   A form of asymmetry in a unimodal frequency distribution. The distribution is positively skewed when more than 50% of elements lie below the mean and negatively skewed when more than 50% of elements lie above the mean (i.e. less than 50% lie below it).

Skewness, measures of.   No measure of skewness is entirely adequate, but as the median, mean and mode of a skew(ed) distribution do not coincide, some useful formulae are:

$$Pe_1 = \frac{\mu - Mo}{\sigma}$$

$$Pe_2 = \frac{\mu - Me}{\sigma}$$

where μ = mean, Mo = mode, Me = median and σ = standard deviation. Pe_1 and Pe_2 are Pearson measures of skewness and these measures can vary between +3 and −3. (◆◆ QUARTILE MEASURE OF SKEWNESS.)

Snedecor's check.   An approximate method of checking standard deviation calculations in normal cases. Briefly, it involves dividing the RANGE by 2 if n (the number of values) is less than 7; by 3 if it lies between 7 and 20; by 4 if it lies between 21 and 40; by 4.5 if it lies between 41 and 80; by 5 if n approximates 100; and so on. As most manual standard deviation calculations are of less than 100 items, this information is all that is needed. For example, the calculation of standard deviation of the series 5,7,10,11,12, using normal conventional methods, would be approximately 2.6; and, if Snedecor's check is used, the procedure would be to determine the range (in this case 7, i.e. 12 − 5) and divide by 2, to obtain 3.5.

In the above case the check is seen to be very approximate. However, the method can be used as a safeguard against obvious errors in quick calculation, such as wrongly multiplying by a factor of 10.

Snowball sample.   A sample in which additional elements (units) are generated by the units first obtained in the sample. If, for example, we wish to obtain a sample of members of a rare population – people suffering from a particular disease, for instance – we might proceed by obtaining, through screening, a number of eligible respondents. We would then obtain from these the names of other sufferers and increase the sample size by using this procedure for each respondent identified. The method is called snowball sampling by analogy with the way in which a snowball (or avalanche) increases in size as it progresses downhill. Snowball sampling is not a probability sampling method but may be useful to obtain a sufficient number of cases for study in certain investigations. Its efficacy depends on the existence of a network of contacts among members of the population being studied.

Spearman's rho (ρ).   A rank CORRELATION COEFFICIENT defined as the PRODUCT MOMENT CORRELATION COEFFICIENT between two sets of rankings of a collection of objects or individuals. Consider the correlation between the rankings 1, 2, 3, 4, 5 and 5, 1, 4, 2, 3.

$$\rho = \frac{\Sigma(x_i - \bar{x})(y_i - \bar{y})}{\sqrt{[\Sigma(x_i - \bar{x})^2]} \times \sqrt{[\Sigma(y_i - \bar{y})^2]}}$$

where x_i is the first ranking of an individual and y_i is the second ranking of the same individual. In this case $\bar{x} = \bar{y} = 3$. Hence

$$\rho = \frac{-3}{10}$$

$$= -0.3$$

The value of the coefficient must lie between -1 (perfect negative correlation) and $+1$ (perfect positive correlation).

A simpler computational formula is

$$\rho = 1 - \frac{6\Sigma d_i^2}{n(n^2 - 1)}$$

where $d_i = x_i - y_i$ and n is the number of paired observations.

Spectral analysis. The analysis of time series by spectral methods measures the strength of various frequency components or periodic terms in the data, and when used to look at the relationship between two or more series it tries to establish how (if at all) that relationship changes between different frequency components. The spectrum is a function that measures how the total variance of a stationary series is distributed between the different frequencies that may be present. It is closely related to the periodogram and indeed the latter is often used as a basis from which to estimate the spectrum (or spectral density function). An alternative but equivalent way of defining the spectrum of a series is known as the Fourier transform of the autocovariance function of the series.

When investigating the relationship between two series we can calculate the cross spectrum which bears the same analogy to the cross-correlation function between two series as does the spectrum to the autocorrelation function. From the cross spectrum we may derive the coherence function which measures the strength of the relationship at different frequencies and the phase function which measures the lead or lag between the series.

Spencer's 15-point moving average. An example of a moving average with unequal weights. It is used to smooth or extract the trend from a time series. The smoothed series at time t is a weighted average of past and future observations with weights that are symmetric. Algebraically a symmetric moving average of length 15 can be written

$$S_t = \sum_{j=-7}^{+7} a_j x_{t-j}$$

where (for symmetry) $a_j = a_{-j}, j = 1, 2, \ldots 7, x_{t-j}$ is the observation in the series at time $t-j$ and S_t is the moving average or smoothed series. In the case of Spencer's 15-point average we have

$a_0 = 74/320$
$a_1 = 67/320$
$a_2 = 46/320$
$a_3 = 21/320$
$a_4 = 3/320$
$a_5 = -15/320$
$a_6 = -6/320$
$a_7 = -3/320$

By adding up these weights it will be seen that their sum plus the sum of $a_{-7} \ldots a_{-1}$, symmetrically defined, is equal to unity. Spencer also proposed a 21-point moving average which is broadly similar in shape and construction but has numerically different weights and three more of them each end.

Splicing.   Economic data often occur with periodic breaks or changes of definition so that we often have to make do with a number of short series based on slightly different definitions. For long-term studies, and especially for use in regression models, restricting attention only to a single definition would leave too sparse a set of data to give useful results. It is often desired, therefore, to link the various parts together into a single series. This can be done either additively or multiplicatively and is known as splicing. By way of illustration consider a price index number that is based on 1968 quantity weights for the period 1968 to 1973 and on 1972 quantity weights from then on. Thus we might have

	1968 based	1972 based	Spliced index
1968	100		100
1969	104		104
1970	112		112
1971	119		119
1972	128	100	128
1973		110	141
1974		125	160

Here the spliced index for 1972 onwards is obtained by multiplying the 1972 based series by a ratio of the index in 1972 based on 1968 = 100 to its value in the 1972 based series (i.e. 128/100). If the overlap is longer than one year we can average the ratios of the 1968 based series to the 1972 based series for the years of overlap. This could be done either as an arithmetic or geometric mean of the overlap ratios. The average ratio can then be applied to the 1972-based series.

SRS.   A commonly used abbreviation for SIMPLE RANDOM SAMPLING.

Standard deviation. The positive square root of the VARIANCE, and one of the most commonly used measures of dispersion. In the case of a NORMAL DISTRIBUTION it is represented by the point on the horizontal axis corresponding to the point of inflexion. As one of the most useful measures of wideness of dispersion of a population it is used in financial planning as a measure of risk. The simple steps in the calculation are:

1. calculate the deviations (differences) between the observations (elements) and their arithmetic mean;
2. find the arithmetic mean of the squared deviations (i.e. the VARIANCE);
3. calculate the square root of the variance.

Example
The following is a population of seven elements:

    6, 10, 15, 19, 23, 27, 33

The arithmetic mean of the observations is 19.

  The differences between the observations and their arithmetic mean are:

    $-13, -9, -4, 0, +4, 8, +14$

The squares of the differences are therefore:

    169, 81, 16, 0, 16, 64, 196

The mean of these differences is:

$$\frac{169 + 81 + 16 + 0 + 16 + 64 + 196}{7} = \frac{542}{7}$$

$$= 77.43$$

The standard deviation is the square root of this value

    $\sqrt{77.43} = 8.80$

This method is adequate for the calculation of the standard deviation of a small ungrouped distribution, and the relevant formula is

$$\sigma = \sqrt{\frac{\Sigma d^2}{N}}$$

where d is the difference between an observation and the arithmetic mean and n is the number of observations.

  Where there are groups of identical observations (elements) (e.g. several '6's or '10's instead of one each, as in the above example) the formula may fully be expressed as

$$\sigma = \sqrt{\frac{\Sigma f_i d_i^2}{N}}$$

where f_i is the frequency of each group (or class) d_i is the deviation of an element in group i from the mean and N is the population size $(=\Sigma f_i)$.

If there is a large number of values an ARBITRARY ORIGIN or ASSUMED MEAN may be used to simplify calculation. The formula is then

$$\sigma = \sqrt{\left[\frac{\Sigma fd'^2}{N} - \left(\frac{\Sigma fd'}{N}\right)^2\right]}$$

where d' is the deviation of an observation from the arbitrary origin.

A number of checks of accuracy of the calculations have been developed. (♦ CHARLIER'S CHECK and SNEDECOR'S CHECK.)

Note: If the standard deviation of a population is being estimated from a sample, then $N - 1$ should be substituted for N in the formulae above.

Standard error.   The standard deviation of the sampling distribution of a statistic, as distinguished from the standard deviation of the population. It is the standard deviation divided by the square root of the number of elements in the sample where sampling is with replacement.

Example

If a population of 1000 men has a mean height of 1.4 m and standard deviation of 0.1 m, the extent of dispersion of samples of size 1 is the standard deviation, i.e. 0.1 m. But suppose that we were to take samples of size 10 from the population. The standard error of the sample mean is given by the formula $\sigma/\sqrt{n}$, which in this case is $0.1/\sqrt{10}$. As the size of the sample increases, so the standard error of the sample mean (i.e. standard deviation of sample mean from the population mean) decreases proportionately to the inverse of the square root of the number of items in the sample. The standard error of the means of samples of size 100 is $0.1/\sqrt{100} = 0.01$.

When comparing means and proportions of two samples, where populations have the standard deviations σ_1 and σ_2 respectively, the standard error of the difference between the sample means is given by the formula

$$\text{s.e.}_{x_1 - x_2} = \sqrt{\left(\frac{\sigma_1^2}{n_1} + \frac{\sigma_2^2}{n_2}\right)}$$

Similarly the standard error of the difference between sample proportions is

$$\text{s.e.}_{p_1 - p_2} = \sqrt{\left(\frac{P_1 Q_1}{n_1} + \frac{P_2 Q_2}{n_2}\right)}$$

where p is the sample proportion, P the population proportion and $Q = 1 - P$.

Standard measure. The transformed measure of a variate by reference to the mean and standard deviation of its distribution, using the function

$$z = \frac{x - \mu}{\sigma},$$

where μ is the mean and σ the standard deviation of the distribution, is called the standard measure.

Example

A value of a variate is 12, the mean is 16 and the standard deviation is 2. As a standard measure the value would be expressed as

$$\frac{12 - 16}{2} = -2$$

(or 2, as an absolute measurement).

When the variate is the sample mean, the measure takes the form

$$\frac{\bar{x} - \mu}{\sigma/\sqrt{n}}$$

where μ is the mean of the population and $\sigma/\sqrt{n}$ is the standard deviation (or standard error) of the sample mean.

Standardization. The expression of a value as a deviate in standard measure, that is, by calculating its difference from the mean of a population and dividing by the standard deviation. The term 'standardization' is also used in sample statistics. If, in regression analysis, the values of dependent and independent variables are standardized the standardized regression coefficients become equivalent to correlation coefficients.

Standardized variable. A variable expressed in STANDARD MEASURE. Variables may be so expressed (1) for comparison with each other and (2) to measure relative distance of observations from their respective means. An illustration of (1) is the expression of examination marks as standardized variables. This would adjust the marks so that differences in the means and standard deviations (dispersion) in different subjects would be removed. Thus the comparison would be in terms of the relative deviations from the respective means.

Stationary process. A STOCHASTIC PROCESS is said to be stationary if its statistical properties are not changing over time. Given any section of the process of length n, the joint distribution of its values must be the same as that for any other section of the same length. In the analysis of time series we are usually content with a much weaker assumption than this, namely that its mean and variance are constant and that the covariance between any two terms depends only on their distance apart in time and not on the particular time to which they refer. Thus

Covariance (x_t, x_{t+s}) = covariance (x_r, x_{r+s})

for all r and t. A process satisfying this weaker condition is said to be second order or weakly stationary.

Statistics. The term originally meant information about a State, or government. Currently it has two meanings, a popular meaning and a specialist one. Popularly it is the plural of 'statistic', and means any information which is presented in numerical form. A wide variety of statistics are readily available, e.g. in the *Annual Abstract of Statistics* and in the *Monthly Digest of Statistics*. In this popular sense the term is almost equivalent to DATA, usually presented in tables or charts.

The secondary and more specialist meaning of the word is the scientific discipline or science of statistics, i.e., the collection of data, the methods by which they can be analysed and tested and the interpretation of the results.

Allied sciences, such as accounting, have their equivalents, i.e. (1) book-keeping, the collection of data; (2) accountancy, its presentation; and (3) management accounting, the interpreting and evaluation of collected data.

Stepwise multiple regression. A method of eliminating or including variables in REGRESSION ANALYSIS by means of a stepwise procedure. The method involves, first of all, the calculation of correlation coefficients between the dependent variable and each of the regressors and the selection of the regressor which has the highest correlation to the dependent variable. The overall F value is then calculated between the regression mean square and the residual mean square. This procedure is explained in VARIANCE, ANALYSIS OF. The correlation matrix is then pivoted, or correlation coefficients calculated between regression estimates using the existing single regressor equation and each of the other regressors, and the regressor which shows the highest correlation with the dependent variable after taking the effect of the first regressor into account is selected.

If the inclusion of the regressor significantly improves the explanation of the dependent variable its regression coefficient is calculated and the coefficient of the earlier regressor recalculated. At each step of the process the correlation matrix is pivoted with respect to the coefficient selected, F-ratio tests are carried out and the DURBIN – WATSON STATISTIC (for AUTOCORRELATION) computed.

Stochastic.   A synonym for chance or probabilistic, as opposed to deterministic. The term originated from marksmanship, the Greek word *stochos* meaning a target. If an arrow is shot at a target, there is an expectation that it will hit it, and the expectation will vary with the skill of the participant. However, there is an element of chance in obtaining the result.

Analogously, in all models there is an expected value or result, but the model is stochastic to the extent that random elements participate in determining the result.

Stochastic model.   As distinguished from DETERMINISTIC MODELS, stochastic models are those which have one or more random components. They may be either: (1) mainly stochastic in content; or (2) partly stochastic, i.e. having a random variable or chance factor in some of the variables.

An example of a model which is mainly stochastic in content is that of the type illustrated under BRANCHING PROCESS and MARKOV PROCESSES, where the model assumes probabilities of division of a future population (or state) into categories.

Other examples of stochastic models are:

(a) multivariate and multiple regression models containing random components;

(b) stochastic LINEAR PROGRAM models; and

(c) stochastic simulation programs, where, as in (b) values used in computing are chosen from a computer program or package which generates random deviates (e.g. Standardized Random Normal Deviates.)

Stochastic process.   A sequence of RANDOM VARIABLES such that the order in which they are recorded is significant. A TIME SERIES is an obvious example.

Strata chart.   A continuous component chart showing the history of a number of components, different categories or items over a period of time as a set of strata. As in the case of a HISTORIGRAM the time scale is measured on the ABSCISSA.

Strata charts differ from COMPONENT BAR CHARTS in showing continuous information, whereas component bar charts may be employed to show discontinuous information of a composite nature.

Example

The chart below shows the values of cash, liquid investments and long-term securities of a banking company between 1970 and 1976.

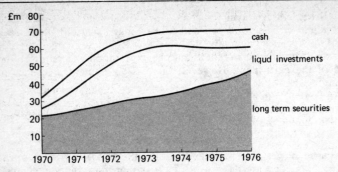

Note that although strata charts usually depict historical information, they may also be occasionally employed to show non-historical continuous composite information.

Stratification. The division of a population into parts, which are called strata. Samples are selected independently within each of the strata. Strata are constructed so that the elements within strata are as homogeneous (similar to one another) as possible and so that the strata are as different to one another as possible. Stratification may be used to increase precision, to guarantee representation of particular groups, to increase the 'representativeness' of the sample and/or to enable us to use different sampling procedures in different strata. Thus in selecting a sample of adults for a survey of attitudes to regional autonomy it would obviously be desirable to ensure that the different regions are represented appropriately in the sample. This can be achieved by treating each region as a stratum, and selecting a sample of appropriate size within each. (◆◆ STRATIFIED SAMPLING, PROPORTIONATE STRATIFICATION, OPTIMAL ALLOCATION).

Stratification after selection. In order to select a stratified sample it is necessary to know to which stratum in the population each unit belongs. This is because for stratification we need to divide the population into parts (known as strata) and then select an independent sample within each stratum – i.e. the stratification is carried out before the sample is selected. In some situations, where we would like to stratify by a particular factor, we may not be able to tell in advance the value of each unit on the stratification factor but we may know the number or proportion of population elements which belong to each category of the stratification factor. In selecting a sample of adults in a city, for instance, we may know the sex distribution of the population but our list (the Electoral Register) does not provide the sex of each individual. The question is whether we can use our information about the actual male/female ratio to improve our estimator.

If we select a simple random sample from the population we can identify the sex of each individual in the sample. We can then treat this sample as a stratified sample with DISPROPORTIONATE ALLOCATION. In this way we can recover most of the gains that could be obtained by proportionate stratified sampling.

Suppose we wish to select a sample of $n = 100$ in a town with a population of 8000 in order to estimate the average earnings of adults in the town. We know in advance that 50% of the adult population of the town is male and 50% female and we expect that sex is related to income. However, we cannot use sex as a stratification factor since our list does not include information on the sex of the individual. In a simple random sample of 100, we may find that we have 45 men and 55 women. The mean income for the men in the sample is £45, say, and for women £30. However, since we know that men and women each account for half the population, we estimate the overall mean income by using the population proportions and not the sample proportions, i.e.

$$\frac{4000}{8000} \times £45 + \frac{4000}{8000} \times £30 = \frac{1}{2}[45 + 30]$$

$$= £37.50$$

This is equivalent to allocating the units to the strata after the sample has been selected and is called stratification after selection, or post-stratification. It is a method of using ancillary information to improve the estimation procedure rather than in the selection process. It may be necessary instead of stratification in three situations:

1. If the values of the stratification variable are not available for classifying and sorting the units before selection.
2. Although the values may be available, they may not have been used.
3. Post-stratification may also be used to improve the estimates for SUBCLASSES where the gains of proportionate stratification may be lost.

The adjustments involved are often slight but are simple to apply and may well be worthwhile.

Stratification factor, stratification variable. When a population is divided up into strata, a criterion or variable used to determine the division is called a stratification factor. We say that we stratify by this factor. For example, if we stratify by sex – i.e. divide the population into two groups, male and female – sex is called the stratification factor or variable.

There can be more than one stratification factor. It may be desirable, for instance, to cross classify the population by both sex and occupation, forming strata which are cells in this cross classification. Thus the group 'female manual workers' could be a single stratum in this case.

Stratified random sampling, stratified sampling.　Stratification implies the division of the population into a number of subgroups, or strata. An attempt is normally made to ensure that each subgroup, or stratum, in the population is constructed in such a way that the individuals or elements within a stratum resemble one another. In stratified random sampling an independent random sample is selected from within each stratum.

For example, if in a sample survey we wish to estimate the attitudes of adults to the Women's Liberation Movement, we might stratify by sex, i.e. categorize the population by males and females and select a sample from each group, rather than a simple sample from the whole population. Since we might expect that men and women would have different attitudes to women's liberation, we ensure by stratifying that both men and women are adequately represented in the sample, and therefore we improve the precision of our estimator.

A second example may help to reinforce this. Before an election the opinion pollsters want to predict the result. To do this they select a small number of constituencies in which to carry out interviews. It is essential, if the prediction is to be accurate, that certain sections of the community should be represented in the sample, e.g. (1) rural areas and urban areas; (2) Labour strongholds; Conservative strongholds and marginal constituencies. Therefore, before selecting the constituencies in which the interviews are to be carried out, the full set of constituencies is divided into these categories and a separate sample is selected from each category, ensuring that each is adequately represented. This procedure should increase the precision of the estimator.

Stratify.　To stratify is to divide the population into relatively homogeneous subpopulations (strata) according to some characteristic of the population elements.

Stratum/strata.　A subpopulation or a section of a population which is designated in advance of the sample selection, and from which a sample is selected independently. The term comes from geology where a stratum is a layer, or set of successive layers, of any deposited substance. In social surveys, strata are frequently formed on the criterion of internal homogeneity, although any identifiable part of the population may be designated as a stratum. (◑ STRATIFICATION, STRATIFIED SAMPLING.)

Structural equation.　Any equation which appears in the explicit formulation of a MODEL. Generally structural models involve a set of equations (relationships) which must be solved simultaneously. In other words, the solution of each equation depends on the solution of the other equations in the model. (◑ PATH MODEL, PATH ANALYSIS.)

Structure. The pattern of relationships between the variables in a MODEL. The structure of a model is concerned only with the pattern and not with the numerical values or coefficients in the model. For example, if we are considering for a population of married women the relationship between age, years of education and number of children born, we might denote the structure diagrammatically as follows:

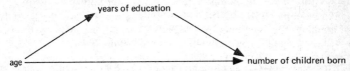

This diagram, which is a PATH DIAGRAM, indicates that the age of a woman is related to the number of children she has; that age is related to the number of years of education (e.g. people now are staying at school longer than they used to); and that education affects family size. More educated women tend to have smaller families.

If the model above were expressed in terms of equations rather than diagrammatically, the equations would be called STRUCTURAL EQUATIONS. It might be preferable to apply the term 'structural variable' only to variables which appear more than once in the system and therefore knit the structure together.

'Student', studentization. Terms associated with the STUDENT'S *t*-TEST and *t*-DISTRIBUTION. The term arose because the original discovery was published in 1908 under the pen-name 'Student'. The writer's identity was disclosed many years later as William Gosset, employed by the Dublin firm of Guinness.

The pen-name has given rise to terms such as Student's hypothesis, Student's distribution, Studentization, etc.

Student's *t*-distribution, *t*-distribution. The distribution of the ratio of a sample mean to a sample variance in samples from a 'normal' population. The distribution is multiplied by a constant in order to make it a probability distribution. It is independent of the parent distribution and can be used to provide confidence intervals of the sample mean independently of the variance of the parent distribution.

It was discovered in 1908 by Gosset, who used the pen-name STUDENT, but later modified by R. A. Fisher in 1925 to the form:

$$dF = \frac{\Gamma[\tfrac{1}{2}(v+1)]}{\Gamma(\tfrac{1}{2}v)\sqrt{(v\pi)}} \left(1 + \frac{t^2}{v}\right)^{-\tfrac{1}{2}(v+1)} dt, \quad -\infty \leqslant t \leqslant +\infty$$

where v is the number of degrees of freedom.

The STUDENT'S t-TEST is based on this distribution.

Student's *t*-test, *t*-test. A test which uses the STUDENT'S *t*-DISTRIBUTION. This distribution expresses the ratio between standard normal and chi-squared distributions and has no reference to the parameters of a population. It may therefore be used when population parameters are unknown.

The most frequent uses of the test are:
1. to compare the means of two small samples;
2. to compare the mean of a small sample with the mean of a population where the population variance is not known.

The CAUCHY DISTRIBUTION is the Student's t-distribution with only one degree of freedom. As the number of DEGREES OF FREEDOM approaches infinity the Student's t-distribution tends to the normal distribution. Tables are available to give the area under the t-distribution curve at given significance levels for different numbers of degrees of freedom. The measurement of the height of the curve, or measure of frequency dF, is given by

$$dF = \frac{\Gamma \frac{1}{2}(\nu+1)}{\Gamma(\frac{1}{2}\nu)\sqrt{(\nu\pi)}} \left(1 + \frac{t^2}{\nu}\right)^{-\frac{1}{2}(\nu+1)} dt, \quad -\infty \leqslant t \leqslant +\infty$$

where ν is the number of degrees of freedom.

Example
In the early part of 1977 Bang Limited had a mean monthly petty cash expenditure of £72. Subsequently, after adjusting values for inflation, the monthly petty cash expenditure was consecutively £85, £77, £92 and £89. There is no apparent change in circumstances. Is there evidence to suggest that expenditure has risen significantly?

Note: the population from which the sample is drawn must be assumed to have a NORMAL DISTRIBUTION.

The t statistic is

$$(\bar{X} - \mu) \times \frac{\sqrt{n}}{s}$$

where μ is the population mean, $\bar{X}$ the sample mean, s the sample standard deviation and n the sample number. The difference between means $(85.75 - 72)$ must be compared with adjusted standard deviation $(s = 6.5)$. Thus $t = (2 \times 13.75/6.5) = 4.2$. At 5% significance level with three degrees of freedom $(n = 4$, so $\nu = n - 1 = 3)$ (using a one-tail test because the hypothesis is simply that $X \geqslant \mu$), the value of t is 2.35. In this case the value of t is 4.2, so the difference is significant. The rise in petty cash expenditure (even after adjustment for inflation) is significant.

Subclass(es).   A subdivision of a class or category in the population or sample being considered. The term is usually used when dealing with a sample. There are two principal kinds of subclass: (1) a cross-class in which the elements are spread over all, or most, of the strata; and (2) a segregated class in which the elements all belong to one, or a very few, of the strata.

Subsample.   A sample of a sample. A subsample may be selected from the original sample for special or intensive treatment ($\blacklozenge$ NON-COVERAGE, MULTIPHASE SAMPLING). It is not necessary that the same method be used to select the subsample as was used in the original sample selection.

Subsampling.   Selecting a sample from a sample. This term is widely used in two different contexts. In MULTISTAGE SAMPLING, the process of selecting second-stage units from within each selected first-stage unit is called subsampling. In MULTIPHASE SAMPLING when a subset of the whole first-phase sample is selected for the more intensive second-phase investigation, the selection process is also called subsampling.

Sufficiency.   The term is used in two contexts: (1) in the context of conditions in logic; and (2) in the context of statistical estimation.
1. This usage is concerned with the common ground between statistics and logic. If all the elements of a population possess the attribute A, then if X belongs to the population 'X has the attribute A'. The condition 'If X belongs to the population' is said to be a sufficient condition for the conclusion 'X has the attribute A'. For example, all living persons have brains, and X is a living person, therefore X has a brain.
2. An estimator of a population parameter is said to be sufficient if the distribution of a sample $x_1, x_2, \ldots x_n$ given the estimator does not depend on the parameter. In effect, the sufficiency of the estimator implies that all information which is relevant to estimation of the parameter is contained in t.

Sum of squares.   The term means essentially the sum of squared values. Thus Σx^2 means the sum of squares of x. However, in statistical terminology the term is sometimes used to indicate the sum of the squared deviations of a set of elements about their mean. If the elements have the values 2, 6, 8, 10, 10 and 12, the sum of squares in this sense can be calculated as

Observations	Deviation from mean	Squared deviations
2	−6	36
6	−2	4
8	0	0
10	+2	4
10	+2	4
12	+4	16
Total = 48 Mean = 48/6 = 8	0	64

Thus the sum of squares in this case is 64. We may note that the VAR-IANCE is the sum of squares divided by the number of elements – in this case 64/6.

The same sense of the term is used in WITHIN-GROUP SUM OF SQUARES and BETWEEN (GROUPS) SUM OF SQUARES.

Supplementary information.   Information on variables other than the survey variables, which may be used either to improve the sample design (for example, by stratification) or to increase the precision of the estimator (for example, in ratio estimation).

Survey.   A study or investigation of a population, usually human beings or economic, social or political institutions. The term may be taken to refer to complete coverage of the population as in a CENSUS, but it is often used to refer to a study dealing only with a sample from the population (i.e. a SAMPLE SURVEY) where the sample observations are used to make inferences or draw conclusions about the whole population.

Survey methods.   All methods, procedures and techniques used in carrying out social surveys are subsumed in the term survey methods. Thus the term covers questionnaire design, sample design, methods of data collection and methods of data analysis.

Survey population.   The POPULATION actually covered by a sample survey. Ideally the survey population should be identical to the TARGET POPULATION for which results are desired, but, mainly due to non-coverage and non-response, there may be differences between the two. Other modifications may be necessary simply to make the survey prac-ticable. For example, individuals living in institutions are frequently excluded from consideration on practical grounds.

Survey sampling.   This deals with methods of selecting a sample from a population in order to use the sample observations, together with information on the method of sampling, to make inferences about the whole population.

The alternative is to carry out a complete enumeration of the popu-lation. Sampling does however have considerable advantages in some cases. (1) Sampling reduces the cost of the study by restricting the enquiry to a (small) proportion of the population elements. The total cost of the study will be greatly reduced, even if the cost per element observed is higher. (2) Sampling reduces the length of time taken to complete the study. This is true not only of the fieldwork but also of the analysis stage. (3) In some situations, sampling is the only feasible approach. Although this is obvious when the observational method **is** destructive – for example, testing explosives in a quality control exercise – it is also true of many social surveys that the skilled personnel necessary to carry out interviews with all the individuals in the population may simply not be available. (4) It may often be true that a sample survey provides a higher level of quality and accuracy than a complete enumeration, since in many situations the quality of the personnel might drop substantially if a compete enumeration were attempted.

A complete enumeration also has some advantages: (a) Information for small groups or parts of the population can be obtained; (b) The public acceptance may be easier to secure, since many people will not be willing to believe that high precision can be obtained with sample data; (c) Sampling statisticians are not needed.

There are two fundamentally different kinds of sampling: (i) PROBABILITY SAMPLING, in which every element in the population has a known non-zero chance of selection; and (ii) MODEL SAMPLING, in which the investigator trusts to his own judgement, or to luck, that 'typical' elements are chosen. The basis of statistical theory is in probability sampling, and other methods should only be used if for some overwhelming reason probability sampling is not possible.

System.   The term system is generally used for something which has in it an element of design, which can be represented by a model. Thus a model is a representation of reality, whereas a system comprises that set of states or processes in reality which the model is designed to represent.

Systematic.   The term means something which is not random. In other cases it may be used even where there is also a stochastic element involved, as in SYSTEMATIC SAMPLING.

Systematic sample.   A sample obtained by the method of SYSTEMATIC SAMPLING. The term is usually confined to samples drawn by selecting units at equally spaced intervals – the SAMPLING INTERVAL. The first sample unit is selected from the first sampling interval by some random method.

Systematic sampling.   A method of selecting a sample which involves taking sampling units from a (numbered) list at equally spaced intervals. The first unit to be selected is determined by some random device (e.g. from a table of RANDOM NUMBERS). This gives the random starting point and the subsequent units are selected by adding the fixed interval to this starting point.

For example, we wish to select five units from a population of twenty units. The length of the SAMPLING INTERVAL is obtained by dividing the population size by the sample size – in this case $20/5 = 4$.

```
     1              ⎛ x 11
     2              |   12          1     2  │  3     4
 x 3             |   13          5     6  │  7     8
 ⎛   4            |   14          9    10  │ 11    12
 |   5            ⎨ x 15         13    14  │ 15    16
 |   6            |   16         17    18  │ 19    20
 ⎨ x 7            |   17
 |   8            |   18                  B
 |   9            ⎝ x 19
 ⎝  10                20
        A
```

One of the numbers from 1 to 4 is selected by some random method. The number 3 could be selected, say. This identifies the first sample unit. The rest of the sample is then identified by adding the interval number of units (in this case, 4) successively to the random starting point (in this case, unit 3) giving the units 7, 11, 15 and 19. The sample of five elements is thus 3, 7, 11, 15 and 19. From Figure B it can be seen that the method of systematic sampling is equivalent here to selecting at random one of the four columns of figure B. Thus, although the units all have equal probabilities of selection, there are only four possible samples – the four columns of Figure B.

Systematic sampling is not equivalent to simple random sampling, unless the list is already in random order. In general, lists are not in random order. There is usually some systematic arrangement involved – alphabetical or seniority, street or house number, for example. Systematic sampling is sometimes called QUASI-RANDOM SAMPLING.

T

t-test. ▶STUDENTS t-TEST

Table (Tabulation). A table is a systematic summary presentation of data. Tables may be:

1. Single-column univariate frequency tables, such as that illustrated under the heading UNIVARIATE, and single column category tables showing the frequencies or relative frequencies of a number of dissimilar categories.

2. Bivariate tables of the type illustrated under MARGINALS or contingency tables as illustrated in CHI-SQUARED TEST.

3. Statistical tables giving critical values of a statistic, such as t, F, χ, ρ, etc. This usage is very common.

Some important points to note in tabulation are:

(a) The headings of the columns and rows should be clear. If the row and column descriptions are too long or unwieldly to use in the table itself, a key may be used, but the full description of the categories should be adjacent to the table.

(b) The data presented should be simple and relevant, and unnecessary detail should be excluded, consistent with the degree of accuracy required.

(c) Detail can be broken at every nth (e.g. 10th) line and tint, or heavy figures, can be used to ensure quick visual extraction of information.

(d) Derived figures, such as averages and percentages, should be placed as near as possible to the originals from which they are calculated.

Target population. The POPULATION about which we wish to obtain information. Every population must be defined in terms of (1) content, (2) units, (3) extent and (4) time. For example, in a study of workers in the manufacturing industry we might define our target population as (a) all workers (b) in firms (c) in Great Britain (d) in 1977. We might have to modify this definition in order to obtain a SURVEY POPULATION with which we could deal in practice. Thus we might define the survey population as (i) all workers (ii) in firms employing five or more workers (iii) in Great Britain (iv) on 1 January 1977.

Target sample. An unsatisfactory term sometimes used to denote the sample originally selected or designated. NON-RESPONSE may cause the achieved sample to fall short of the target sample.

Tchebycheff's inequality. For any distribution with a finite standard deviation, Tchebycheff's inequality gives an upper bound (or upper limit) to the proportion of the distribution lying more than any fixed number of standard deviations from the mean. In other words, the inequality gives an upper limit to the probability of a single observation, taken at random, falling outside the fixed interval.

Thus, if we write the true mean as μ and the standard deviation as σ

$$P\{|x - \mu| \geq \lambda\sigma\} \leq 1/\lambda^2$$

for any value of λ.

In most practical cases, this inequality is very conservative.

Telescoping. A type of memory or RECALL ERROR. When respondents are asked to give information about the past, the question usually involves a reference period – for example, 'How much money did you spend on household repairs in the last three months?' Frequently the respondent may include in his answer, by mistake, items of expenditure which occurred more than three months previously, particularly if the expenditure was large. This process is called TELESCOPING. If the question were phrased as 'How much money did you spend on household repairs between June and September last?' the reverse process might also occur – items of expenditure since September might also be included in the response. The method of BOUNDED RECALL may be used to deal with this problem.

Test statistic. A function of a set of observations from a sample, which acts as a basis for the testing of a hypothesis. Thus, for a given set of sample observations the *Student's t*-STATISTIC, and the Chi-Squared statistic, for example, are test statistics. (◆◆ STUDENT'S *t*- TEST, CHI-SQUARED TEST.)

Three-stage sample. A widely used special case of MULTISTAGE SAMPLING. The population is assumed to be classified into first-stage units (PRIMARY SAMPLING UNITS). In selecting a sample of adults in Great Britain these first stage units might be the parliamentary constituencies. A sample of first-stage units is selected and each of the selected units is divided into second-stage units (secondary sampling units). In Britain these second-stage units might be polling districts. A sample of second-stage units is selected within each selected first-stage unit. Finally the selected second stage units are divided into final-stage units and a sample of these is selected from each. In the example above the final-stage units would be the individual electors. Thus the full procedure would be: (1) to select a sample of constituencies – the first stage;

(2) to select a sample of polling districts within each selected constituency – the second stage; (3) to select a sample of individuals within each selected polling district – the third stage. The method can be applied to any population in which suitable units are available. In a sample of industrial workers, for example, the first stage units might be firms; the second stage units departments within firms; and the third stage units the individual workers.

Tied ranks. In ranking procedures, two or more objects or persons may achieve levels of success (or failure) which are indistinguishable from each other. In this case, it is usual, when calculating rank correlation coefficients, to allot ranks to each equally successful constituent of the group equivalent to the mean of the ranks which they would have received if they were not equally successful.

Example
The marks of ten children in an examination are

   84, 70, 65, 65, 60, 57, 57, 57, 40, 32.

In this example, two sets of ranks are tied at 65 and 57 respectively and they cannot therefore be ranked sequentially from 1 to 10. Instead the third and fourth candidates are allocated the mean of 3 and 4 (= 3½) and the sixth, seventh and eighth candidates are allocated the mean rank of 7. Thus the ranks of the 10 children are

   1, 2, 3½, 3½, 5, 7, 7, 7, 9, 10.

Time reversal test. For a single commodity we can express (say) price in year 1 compared to year 0 as the ratio (p_1/p_0), which when multiplied by 100 gives an index of year 1 based on year 0 = 100. Conversely the index of year 0 based on year 1 (= 100) is $100 \times p_0/p_1$. In other words, reversing the direction of time produces (apart from the factor 100) the reciprocal of the index. It is sometimes required that this property should extend to an index number of many prices. An index $I_t^{(0)}$ for year t for year t based on year 0 = 100 satisfies the time reversal test if

$$\frac{I_t^{(0)}}{100} \times \frac{I_0^{(t)}}{100} = 1$$

Neither the Laspeyre nor the Paasche indices have this property although Fisher's 'ideal' index does have it. Consider, for example, the Paasche index and apply the time reversal test

$$\frac{\Sigma p_t q_t}{\Sigma p_0 q_t} \times \frac{\Sigma p_0 q_0}{\Sigma p_t q_0} \neq 1$$

i.e. the Paasche index does not satisfy the test.

Time sampling. The sampling of observations of a variate over a period of time, as distinct from the extraction of a sample from a population at a given point of time. Contrast, for example, the sampling of temperature measurements at a particular place on each day for a number of years (time sampling) with the sampling of weights of children at a school on a given day.

Time sequence. A set or array of variate values ordered chronologically (i.e. in time order) as distinct from a TIME SERIES, where a variate is observed consistently at fixed intervals of time.

Time series. A set of observations on a random variable made at successive points in time, or indeed any ordered set of measurements of a random variable, is called a time series. It is important that the ordering has some significance if the data are to be analysed by time series methods. The point about analysing time series is that we believe the time dimension can help us to understand the causal mechanism generating the data or at least enable us to recognize, and use to our advantage, the local patterns and co-ordination observed in the data. Examples are multitudinous. Nearly all economic variables are recorded monthly, quarterly or annually, and time series data also occur frequently in such diverse fields as astronomy, geology, meteorology, chemical engineering, biology and many others.

When analysing data of this type we must must first postulate a model generating it, the components of which we (may wish to) isolate. A fairly simple model from a series $\{X_t\}$ would be:

$$X_t = T_t + C_t + S_t + I_t$$

where T_t is the trend component at time t, C_t a cyclical (business cycle) component, S_t a seasonal component, and I_t an erratic or random component. Seasonal adjustment, for example, would consist of estimating and then subtracting the term S_t. Our interest in C_t might be to pick out its turning points. Other models assume that X_t is made up of a (possibly large) number of cyclical components whose amplitudes and frequencies must be estimated (spectral analysis), or that X_t is generated from its own past values and some disturbance terms (autoregressive models).

Tolerance. A term used in statistical QUALITY CONTROL to mean the variation from a given norm or standard which is acceptable for the purpose of production, the standard or norm being expressed as the mean of the group. Thus tolerance limits, for example, are those limits within which an article's size, or other measure of component input or output must lie if the article is to be acceptable. In practice there are two sets of such limits, warning limits and action limits. If a single article in a sample has a size or other relevant measurement (i.e. component input or output) which is outside a pre-stated action limit, the batch is carefully examined, and if a group of consecutive articles have measurements falling between warning and action limits the criteria of production and the actual process are carefully examined.

Sometimes the term 'tolerance factor' is used, but this is variously defined, for example

$$\frac{T_U - T_L}{\sigma}$$

where T_U and T_L are respectively the upper and lower tolerance limits and where σ is the known standard deviation in the measurements of the product size, component input quantity or other output factor which can be regarded as acceptable. In the above formula the tolerance limits, strictly interpreted, are the action limits, that is, they are the pair of limits farthest removed from the mean, or recommended standard of measurement.

Total correlation. In multiple regression analysis and multivariate analysis, total correlation is the raw correlation between two variables irrespective of the influence of a common factor. It is to be distinguished from (1) the correlation of each of the variables with other(s) representing common factor(s) and (2) the correlation between residuals after the common factors have been extracted.

Total effect. The total effect of a variable on another in a PATH MODEL is the sum of the DIRECT EFFECT and INDIRECT EFFECT. Thus it represents all the influence one variable has on another whether it operates directly or through its impact on some intermediate variable in the model. For example, the effect of mother's education on family size in a model including marital duration can operate in two ways; first, the effect that additional education has in delaying entry into marriage and therefore by reducing effective marital duration reduces family size – the indirect effect; second, the direct effect of education in producing a different attitude to number of children wanted – the direct effect. Diagrammatically, this is represented as

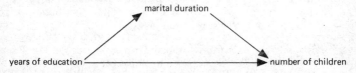

Trend. In a time series, the trend is the broad underlying movement. However, it must be seen in relation to the period of observation available for study. Two or three years data on the upswing of a business cycle would appear as an upward trend, but if extended further the cyclical pattern causing this apparent trend would emerge.

We would need several complete cycles to be able to distinguish properly the trend and cyclical terms, and even when we have done this we can never really be sure the trend is not a reflection of a much longer cycle. Our definition of trend then should depend also on the use we may make of it and on how far ahead we asssume explicitly or implicitly that it will continue. Bearing in mind the above reservations, the trend in a time series can be estimated either by fitting an appropriate mathematical function using regression methods, or by smoothing the cyclical and erractic components out using a moving average.

Trend free series.   This is a series in which no trend is present either because the variable concerned historically has fluctuated about a constant level, or because whatever trend was originally in the data has been removed by statistical adjustment.

Truncation.   The word is used in two senses:
1. The omission from a population of all values above or below a fixed value. For example, if the majority of values in a population are below 100 and there are reasonable grounds for believing that values over 100 are not important, all values over 100 may be ignored.
2. The omission of a decimal fraction from a value (as opposed to rounding). For example, the number 15.98 is rounded to 16 as its nearest integer, but when truncated it is expressed as 15.

Two-phase sampling.   A sampling procedure in which the final sample is not selected directly from the population but from the first-phase sample, through which information is obtained to improve either the selection or the estimation or both. (◆ MULTIPHASE SAMPLING.)

Two-stage sampling.   A simple case of MULTISTAGE SAMPLING. The population is classified into groups of elements – the primary sampling units. A sample of these primary sampling units is selected – the first stage; the selected primary units are then classified into the basic sampling units – the secondary sampling units; and a subsample of secondary units is then selected from each – this constitutes the second stage.

Two-tailed test.   A test where the area of rejection comprises two areas, one at each extreme (or tail) of the sampling distribution curve relevant to the test under consideration.

Example.
1. If it is desired simply to test for significant overusage of a material (for quantity control purposes), a standard usage will be determined and, if usage exceeds a given measure of difference (e.g. 1.640 standard errors or 2.330 standard errors) more than the given standard some action will be taken. In this case, no action need be taken for underusage, so that a single-tailed test would be satisfactory.
2. If, in production, it is desired to test for quality and determine

whether screws are being produced which are either too long or too short, then the region of rejection will include two areas of unusable size, those smaller than a stated minimum and those larger than a stated maximum. This kind of test would involve the two tails (or extremes) of the normal distribution curve. A double-tailed test would be necessary.

Type I error(s). An error involving the rejection of the null hypothesis when it is true.

Type II error(s). An error involving the acceptance of the null hypothesis when it is false.

U

Ultimate cluster. A term sometimes used to denote the set of elements from a particular primary sampling unit which are included in the sample. In the example given above of MULTISTAGE SAMPLING, the ultimate clusters are the sets of individuals eventually selected in each of the constituencies.

Unbiased. A sample, estimation procedure or estimator which is free from BIAS is said to be unbiased. Thus no systematic distortion or error is present in the procedure. Such a systematic error should be distinguished from a random error, which may produce a distortion on any particular occasion but balances out (cancels out) on the average.

Thus if an estimator takes the values 3, 7, 12, 2, 15 and 9 each with relative frequency or probability 1/6, the estimator is an unbiased estimator of a parameter with value 8. The errors are -5, -1, $+4$, -6, $+7$ and $+1$ which cancel out over all samples.

Unbiased error(s). An error, i.e. a deviation between the true and observed values, which has an expected value of zero over repeated sampling or repeated observation. Therefore, although individual errors may be positive or negative, in the long run these errors will tend to cancel out. Total error can be divided into two components – the systematic error or BIAS and the random errors which have an expected mean of zero, the unbiased errors.

Unbiased estimator. If an estimator has an EXPECTED VALUE, for all sample sizes, equal to the parameter it is designed to estimate, it is an unbiased estimator. Thus in repeated sampling, the estimator takes on average the true value of the parameter. Therefore although for any particular sample the estimate obtained may not be correct (i.e. may not be equal to the parameter being estimated), the average taken over all possible samples will be equal to the parameter. The estimator is thus free from systematic error or distortion. (◊ UNBIASED.)

Unbiasedness. The property of being free from BIAS.

Unidimensionality. The capacity to be measured in only one dimension. A univariate frequency array, for example, is a one-dimension array. Although two axes are used for the purpose of showing frequency values, in practice a single variate could be depicted along a single axis or by using a single vector of frequencies.

The term is also used widely in attitude measurement. Unidimensionality in this context means that all of the items in a particular scale refer to the same attitudinal dimension.

Uniform sampling fraction.   In STRATIFIED SAMPLING, if the number of units selected from each stratum is proportional to the number of units in that stratum, the sample is said to be selected with a uniform sampling fraction.

If we wish to select with a uniform sampling fraction a sample of 1000 from a population of 1 000 000 which is divided into three strata of sizes 500 000, 300 000 and 200 000 respectively, we would proceed as follows:
1. Overall sampling fraction = 1000/1 000 000 = 1/1000.
2. In stratum 1, applying this sampling fraction gives a stratum sample size of 500 000/1000 = 500.
3. In stratum 2, applying the same sampling fraction gives a stratum sample of 300 000/1000 = 300.
4. In stratum 3, applying the same sampling fraction again gives a stratum sample size of 200 000/1000 = 200.

Thus the overall sampling fraction 1000/1 000 000 is equal to the sampling fractions in each of the strata: 500/500 000, 300/300 000, 200/200 000.

(◆◆ PROPORTIONATE STRATIFIED SAMPLING.)

Uniformly better estimator.   In comparing two estimators T_1 and T_2 of a population parameter, if, using a particular criterion, T_1 is at least as good as T_2 at all possible parameter values, and better at some parameter values, we say that T_1 is uniformly better than T_2.

Uniformly most powerful.   For tests with a given probability of type I error (or a test of size) α, a test which has a power on the alternative hypothesis no less than any other test of size α is called a uniformly most powerful (UMP) test.

Unimodal.   A term used to describe a frequency distribution which has only one MODE. For example, the NORMAL DISTRIBUTION has only one mode, which coincides with its mean and median, and ASYMMETRIC or SKEWED DISTRIBUTIONS may also have single modes which do not coincide with their means and medians. (◆◆ MEAN, MEDIAN.)

Unit.   A means of measuring or counting data for a particular purpose, or a homogeneous item in which data are either counted or measured. It may be a unit of measurement, such as a kilometre, pound, dollar or hour, or a unit counted for measuring frequency, such as a person, a brick, a share or even an examination mark.

Some characteristics of statistical units are that they must be:

1. *Specific.* It is not of much use to add loaves of bread to packets of cheese for the compilation of a cost of living index. Items counted must not only be of the same kind, but capable of aggregation without any bias because of size, gender or type.

2. *Stable.* Constituents of data must not change over time. If there is a change, allowance must be made for it. For example, a firm's total profits may be computed over a number of years on the assumption that the pound is stable. If inflation occurs, it is preferable to convert past values of the pound sterling to current values for the purpose of aggregation.

3. *Relevant.* The unit of measurement must be appropriate to the data being studied.

It follows from these three characteristics that units should be homogeneous (for they must be specific in type and stable over time). Sometimes artificial homogeneity is achieved. For example, prices from different periods may be compared by means of index construction and different kinds of work may be compared by using standard hours. Education statistics are sometimes measured in student hours and hospital statistics in patient-beds or in patient-days. These are all attempts to find a statistical unit (often a composite one) which has the three essential characteristics.

Unit of analysis.   The subdivision of the population which forms the basis of the analysis. There may well be different units of analysis within the same survey or study. In a study of expenditure patterns, for example, some of the analysis may be carried out using the individual as the unit of analysis; other analyses may be based on the expenditure of households – here the unit of analysis is the household.

Univariate (univariate distribution).   The term univariate simply means having only one variate (or variable). A univariate distribution is a single distribution consisting of only one variate, which may be either continuous or discontinuous. For example, the heights of a set of children can be represented in the following table:

Height (cm)	Frequency
Under 80	2
80 and under 100	10
100 and under 120	27
120 and under 140	39
140 and under 160	28
160 and under 180	9
180 and over	3

In contrast are bivariate (two-variate), trivariate (three-variate), etc. distributions.

Unrestricted random sampling.   A sampling method in which every element in the population has an equal probability of selection *at each drawing*. Thus, if we wish to select a sample of size $n = 2$ from a population of size $N = 4$, each element has an equal probability $(1/4)$ of being selected at each of the two drawings. Thus the possible samples from the population $\{1, 2, 3, 4\}$ are:

$\{1, 1\}$	$\{2, 1\}$	$\{3, 1\}$	$\{4, 1\}$
$\{1, 2\}$	$\{2, 2\}$	$\{3, 2\}$	$\{4, 2\}$
$\{1, 3\}$	$\{2, 3\}$	$\{3, 3\}$	$\{4, 3\}$
$\{1, 4\}$	$\{2, 4\}$	$\{3, 4\}$	$\{4, 4\}$

Each element appears an equal number of times in these samples and has a probability of 1/2 of being selected. Unrestricted random sampling is less precise than simple random sampling because of the possibility of an element appearing more than once in the sample – no additional information is obtained by observing or measuring that element more than once. However, estimators based on unrestricted random sampling have attractive mathematical properties and most statistical theory is based on the assumption of unrestricted random sampling. In practice in sample surveys, sampling is almost always carried out without replacement.

The term 'simple sampling' is sometimes used as a synonym for unrestricted random sampling. (◗◗ SIMPLE RANDOM SAMPLING.)

Universe.   A term meaning POPULATION or parent group from which a sample is taken. The term 'universe' is older than population, and related originally in logic to the Aristotelian concept of the universal set, that is, the *general*, as distinct from the *particular*, e.g. *all* brown-eyed girls, *all* flat-nosed Manx cats, etc. However, since each universe is, by its very nature, a subset of something else, as *all* brown-eyed girls are a subset of *all* girls or *all* flat-nosed Manx cats a subset of *all* cats, the term is less common in practical statistics than it was in classical logic. It is best to define each population in relationship to its context.

If all the share movements of the United Kingdom are being studied, for example, the universe from which the sample is taken consists of all shares. If only textile shares are being studied, the universe is restricted to all textile shares.

V

Valid, validity. A scale is valid if differences between the scale scores represent true differences in the characteristic which we are trying to measure. Thus the validity of a scale depends on the degree to which it succeeds in measuring the characteristic under study. Validity cannot ever be assessed from the data alone; some external source or subjective evaluation is necessary.

The measurement of validity poses serious problems. At the most superficial level, the validity of an attitude scale may be examined simply by checking that all the items in the scale are dealing with the attitudinal continuum (e.g. attitude to other races) under consideration. This is FACE VALIDITY. Other approaches are the examination of content validity, predictive validity and CONSTRUCT VALIDITY.

Validation. A procedure which uses external sources independent of the investigation in order to ascertain whether the measurement process is free from bias. Checking the responses obtained in an interview against official records is an example of a validation exercise. The term is also used to denote a check carried out to find out whether a sample is representative of the population from which it was drawn.

Variable. Any quantity which varies in measurement. For example, the height of one person is different from that of another. Similarly the weight of one person differs from that of another. Quantities which vary in value and have an associated probability distribution are known as RANDOM VARIABLES, or VARIATES. The term variable is sometimes used loosely as a substitute for variate.

Variance. The mean of the sum of squares of all differences between values of a variate and the overall mean of all values. The STANDARD DEVIATION is the square root of the variance.

Example
A variable has the following values

6, 10, 15, 19, 23, 27, 33

The arithmetic mean of all values is 19.

The differences between each of the values and the arithmetic mean are

$$-13, -9, -4, 0, +4, +8, +14$$

The squares of the differences are therefore

$$169, 81, 16, 0, 16, 64, 196$$

The variance is the mean of these differences

$$\frac{169 + 81 + 16 + 0 + 16 + 64 + 196}{7} = \frac{542}{7}$$

$$= 77.43$$

The variance is a valuable measure in a PROBABILITY DISTRIBUTION and is one of the most widely used measures of dispersion. It may be partitioned to determine the sources of variation. (◆◆ VARIANCE, ANALYSIS OF.)

Variance, analysis of. A method of dividing the total variation of observations into components which can be attributed to or associated with particular sources of variation, e.g., the difference between groups or classes used in classifying the observations.

Example
Bert, Jock, Sidney, Seamus and Taff are partners in a subcontracting firm sharing profits equally. The four partners other than Sidney believe that he is not pulling his weight. Sidney argues that there is so much difference in work done on different days of the week that it would be difficult for the other four partners to prove their point. In a sample week, the following figures are available.

	Jobs performed (assumed to be equal tasks)				
	Bert	Jock	Sidney	Seamus	Taff
Monday	6	8	1	15	11
Tuesday	8	10	2	16	12
Wednesday	8	10	2	18	14
Thursday	9	11	3	15	13
Friday	9	11	2	21	15
Totals	40	50	10	85	65
Means	8	10	2	17	13

The grand mean for all the partners is ten jobs per day.

What is the evidence that there is a significant difference between the performance of each of the partners and the performance of the partners on different days on the week?

First, let us calculate the total sum of squares. This means subtracting the grand mean from each performance statistic and squaring the difference.

| | \multicolumn{5}{c}{Squares of differences} |||||
	Bert	Jock	Sidney	Seamus	Taff
Monday	16	4	81	25	1
Tuesday	4	0	64	36	4
Wednesday	4	0	64	64	16
Thursday	1	1	49	25	9
Friday	1	1	64	121	25

The total sum of squares is 680, and as the DEGREES OF FREEDOM are 24 (i.e. 25− 1), the estimate of the population variance, or mean square, from this sample is

$$\frac{680}{24} = 28.\overline{3}$$

If we assume that there is no interaction, we may now calculate the 'between partners sum of squares' by squaring the differences between the partners' means and the grand mean (= 10), multiplying by 5 and aggregating. Thus the between partners sum of squares is

$$5 \times (2^2 + 0^2 + 8^2 + 7^2 + 3^2) = 630$$

and as the degrees of freedom are 4(i.e. 5 − 1), the mean square is

$$\frac{630}{4} = 157.50$$

The following table shows the components into which the sum of squares may be divided, using the above method.

Component	Sum of squares	DF	Mean square	F-values
Between partners	630.0	4	157.50	131.25
Between days	30.8	4	7.70	6.42
Residual	19.2	16	1.20	
Total	680.0	24	28.3$\overline{3}$	

There are 4 degrees of freedom in the numerator and 16 in the denominator. *F*-value tables are two-way tables constructed to show specific *F*-values given *m* degrees of freedom in the numerator and *n* degrees of freedom in the denominator.

In this case the critical *F*-value at the 1% significance level is 4.77. Both the 'between partners' and the 'between days' *F*-values are therefore significant.

Variate. A quantity which varies with a given frequency distribution. The term is synonymous with RANDOM VARIABLE. (◗ VARIABLE.)

Vector. A matrix with either only one row or one column, i.e., $k \times 1$ or $1 \times k$ matrix.

Thus

$$\begin{pmatrix} 1 \\ 9 \\ 6 \end{pmatrix}$$

is a column vector; and

$$(1 \quad 9 \quad 6)$$

is a row vector.

Venn diagram. This is used in set theory to illustrate probabilities. Imagine a total set of twenty-four persons, of which twelve are brown-eyed and eight are female. There are only three brown-eyed females in the set. The probabilities of taking at random a brown-eyed male, a non-brown-eyed female, a female or a male from such a set may be illustrated, using a Venn diagram.

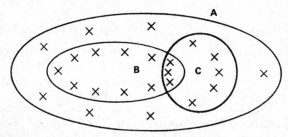

In the above diagram, the large ellipse *A* depicts the total or universal set of twenty-four persons, the small ellipse *B* depicts the set of twelve brown-eyed persons, and the circle *C* depicts the set of females. The intersection of the circle and the small ellipse depicts the subset of brown-eyed females. By examining the diagram, and without any knowledge of the mathematical rules of, for instance, conditional probability, it is possible to calculate that the probability of selecting:

A brown-eyed male is $\dfrac{9}{24} = 0.375$ $(B \wedge \sim C)$

A non-brown-eyed female is $\dfrac{5}{24} = 0.208$ $(\sim B \wedge C)$

A female is $\dfrac{8}{24} = 0.333$ (C)

A male is $\dfrac{16}{24} = 0.667$ $(\sim C)$

where $\sim C$ denotes the complement of C i.e. all the elements not in C.

Thus probabilities can be calculated simply, without direct application of the rules of probability.

W

Weight. A number by which a value is multiplied in order to give it a required measure of importance in relationship to other values. The term is used in simple statistics in the calculation of index numbers and averages and in sampling. For example, if it is considered that three major means of subsistence – bread, cheese and tea – have the relative measures of importance 3, 2 and 1 respectively, then the prices of all three may be used to compare the cost of subsistence in year 0 with that in year 1 by multiplying the prices of each of the commodities in years 0 and 1 with their respective 'weights' 3, 2 and 1 and comparing the aggregates of the products.

Example

Commodity	'Weight'	Price		Product	
		Year 0	Year 1	Year 0	Year 1
Bread	3	25p	30p	75	90
Cheese	2	15p	18p	30	36
Tea	1	10p	12p	10	12
				115	138

The aggregate products of prices and 'weights' may be compared either by compilation of an index [138/115 × 100 = 120] or by using the absolute difference between the products for each year in some other way.

Weights are also used in SAMPLING to achieve proportionality, as, for example, in stratified sampling when there is a difference in the probability of selection of elements from a number of different strata. In order that the total sample consisting of all strata be made representative, weights are assigned to the sample means for the strata, each of the weights being the inverse of the element's probability of selection.

Weighted average.   A mean calculated from values of unequal importance. The term is used in three contexts: (1) in calculating an
estimator of a population mean from a sample selected with unequal
probabilities, where each weight used is based on the inverse of the
probability of selection; (2) in combining the means of strata samples
in stratified sampling; and (3) in business calculations to calculate
average profit for goodwill purposes or to compute average inventory
prices.

Example
In business calculations, recent profits are better indicators of current
value than profits of previous years, and it is common practice to
'weight' the profits of five or more years using relative factors 1, 2, 3, 4
and 5. These weights are entirely arbitrary but take into account the
relevance of the more recent years.

As a worked example let us calculate the weighted average profit of
Utils Limited for the years 1972 to 1976, from the consecutive profit
figures £10 000, £12 000, £16 000, £13 000 and £15 000.

Year	Profit (£)	'Weight'	Product (£)
1972	10 000	1	10 000
1973	12 000	2	24 000
1974	16 000	3	48 000
1975	13 000	4	52 000
1976	15 000	5	75 000
Total 'weights'		15	£209 000

Thus

$$\text{Weighted average of profits is } £\left(\frac{209\ 000}{15}\right) = £13\ 933$$

The same method is used for calculating the weighted mean when
sampling units of unequal importance and in stratified sampling; in
such cases, though, the weights are not arbitrary, each weight being
derived from the inverse of the probability of selection. (◆ WEIGHT.)

**Wilcoxon's matched pairs signed ranks test, Wilcoxon's signed ranks
test.**   A test which was devised by Wilcoxon in 1945 to compare two
random samples of matched measurements by reference to the sign
(i.e. + or −) of each of the differences between them. The test is more
powerful than WILCOXON'S SUM OF RANKS TEST since it takes into
account both the *direction* (+ or −) and the *size* of differences between

matched pairs, and it also has the feature that, if there is no significant difference between the two groups, the numbers of positive and negative differences should be similar. Probability is assessed by using tables of measurements which are available for use where the number of pairs ranges between 6 and 25. For higher numbers of pairs the equivalent of the standard normal score (z) may be estimated using the formula

$$z = \frac{T - \mu_T}{\sigma_T}$$

For ranked values this is

$$z = \frac{T - [N(N + 1)/4]}{\sqrt{\{[N(N + 1)(2N + 1)]/24\}}}$$

where N is the number of pairs and T is the total of ranks with the less frequent sign.

For practical purposes this formula may be simplified to

$$z = \frac{2T - \tfrac{1}{2}N(N + 1)}{\sqrt{\{[N(N + 1)(2N + 1)]/6\}}}$$

For this test, as with WILCOXON'S SUM OF RANKS TEST, tables are available which show the minimum critical values of T for any given number of pairs (N).

Example

Two lecturers, Parsons and Samuels, are being interviewed for appointment at a college, and the appointing board takes the rather unusual course of asking each lecturer to talk for 20 minutes to a set of ten randomly picked students on a subject with which the students are unfamiliar. The group is then tested a while afterwards with similar-sized tests, containing thirty questions on each subject. The following table shows their marks.

lecturer	subject / students	1	2	3	4	5	6	7	8	9	10
Parsons	parameters	21	28	18	19	22	27	10	17	19	20
Samuels	samples	23	25	15	20	19	25	14	18	16	17
differences		−2	+3	+3	−1	+3	+2	−4	−1	+3	+3
ranks of differences		3½	7	7	1½	7	3½	10	1½	7	7

The table below shows the frequencies of positive and negative ranks of differences. The method of ranking is explained under TIED RANKS.

differences	ranks of differences	+ ranks	− ranks
1	1, 2, (2 × 1½)		3
2	3, 4, (2 × 3½)	3½	3½
3	5, 6, 7, 8, 9 (5 × 7)	35	
4	10 (1 × 10)		10
	(T) totals of ranks	38½	16½

In this test, a significant difference between pairs is indicated if the smaller value of T is less than the critical value at a given significance level for a particular number (N) of pairs. In the example, the smaller value of T is 16½. Tables show that where $N = 10$, the smaller value of T may be as low as 10½ at a significance level at 10%, as low as 8 at a significance level of 5%, and as low as 5 at a significance level of 1%. Although the results seem to suggest that Parsons is a better lecturer than Samuels because there are more positive differences than negative ones, the test results do not indicate that there is a significant difference between the two lecturers.

Wilcoxon's sum of ranks test. A distribution-free method of comparing two unmatched random samples of measurements by using a ranking procedure, pooling the measurements and employing a system of paired comparison. Although the test can often be used in the same context as the STUDENT'S t-TEST it does not require the assumption of normality, and is more robust, but less powerful in the normal case, because it relies on the ranks of values, rather than the values themselves.

Example
The following table shows the amount of time in minutes which candidates from two schools take to obtain the answer to an examination question.

Candidate	School X (min)	Candidate	School Y (min)
1	21	10	29
2	24	11	33
3	24	12	22
4	30	13	31
5	23	14	28
6	22	15	38
7	27	16	29
8	28	17	24
9	25	18	29

Do these results differ significantly from each other? If we were to compare the means of results for both schools we should find a difference of about 5 min in the mean time taken to answer the examination question. Does this mean that school Y is significantly slower than school X?

Procedure

Prepare a table showing the values, the overall rank of each of the values and the number of candidates from each group to achieve the ranks, using tied ranks for equalities.

														Total
Frequency	1	2	1	3	1	1	2	3	1	1	1	1	18
Values (min)	21	22	23	24	25	27	28	29	30	31	33	38	
Tied ranks	1	2½	4	6	8	9	10½	13	15	16	17	18	
School X	1	2½	4	12	8	9	10½	15					62
School Y		2½		6			10½	39		16	17	18	109

Although the two totals of ranks are given for the paired comparison, the important value is the smaller total (school X = 62). Tables for Wilcoxon's sum of ranks test show minimum critical values (in contrast with most other tests where maximum values are given). Critical values are just as low as 62 where $P = 5\%$, and 66 where $P = 10\%$. The difference is therefore only just significant at 5% significant level, but is significant at the 10% significance level.

Within-group sum of squares. In ANALYSIS OF VARIANCE the total sum of squares of the observations from the overall mean can be broken down into two components: (1) the BETWEEN GROUPS SUM OF SQUARES and (2) the 'within group sum of squares'. The latter is the total of the sums of squares of the observations in each group about the group mean.

Y

Yates correction, Yates modified chi-square test. One of the modifications of the CHI-SQUARE TEST used on 2 × 2 contingency tables, where the expected number of occurrences in each of the four subcategories is more than 5 and where the total number of actual occurrences in all four categories is more than 50. Although the unmodified chi-square test can be used successfully on contingency tables of 3 × 2 subcategories, a modification is necessary in contingency tables of 2 × 2 subcategories where there is only one degree of freedom. The correction is to subtract 0.5 from each value of |O − E|, effectively decreasing the observed values where O > E and increasingly observed values where O < E, by 0.5 in each case.

Example
In an examination, passes and failures, male and female are categorized.

	Male	Female	Total
Pass	300	140	440
Failure	200	160	360
Total	500	300	800

The expected values using the criteria stated in the CHI-SQUARE TEST are Male Pass 275; Male Failure 225, Female Pass 165; and Female Failure 135, so that the value |O − E| in all four cases is 25. This value is modified to $(25 - 0.5) = 24.5$ so that $|O - E|^2$ is revalued at $(24.5)^2 (= 600.25)$ instead of $(25)^2 (= 625)$.

$$\text{Thus } \chi^2 = \frac{600.25}{275} + \frac{600.25}{225} + \frac{600.25}{165} + \frac{600.25}{135}$$

$$= 2.18 + 2.70 + 3.64 + 4.45 = 12.97$$

For the purposes of Yates correction at 1 degree of freedom, the values of χ^2 are 2.71 where P = 10%, 3.84 where P = 5% and 6.63 where P = 1%. The value of 12.97 is thus highly significant and the probability of such a result arising purely by chance is less than 1%.

Z

Z-score statistic. The expression of the value of an observation using one standard deviation as the unit of measurement of its distance (i.e. positive or negative difference) from the mean of the distribution of the variate.

Example
If 3.9 is a given value of a variate whose mean is 9.3 and whose standard deviation is 2.7, the Z-score statistic of the value is

$$\frac{3.9 - 9.3}{2.7} = -2$$

Z-score statistics serve to show the extent to which a particular value of a variate differs from the mean of the distribution of the variate.

Z-transformation of the correlation coefficient. A transformation of the correlation coefficient using the formula

$$z = \tanh^{-1} r$$

and taking into account the fact that the distribution of z for samples from a bivariate normal population approaches the normal distribution more quickly than does r.

It should be noted that where the number of pairs, n, in the bivariate sample exceeds 36, the critical value of the correlation coefficient for a given significance level may be approximated by

$$r_c = \frac{z}{\sqrt{(n-1)}}$$

where r_c is the critical value of the correlation coefficient; and z is the critical value of z for the normal distribution. Thus where there are 37 pairs of observations at the 5% significance level the critical value of the correlation coefficient is

$$\frac{1.96}{\sqrt{36}} = 0.33$$

Zee chart (or Z chart). A graphical depiction showing single values, progressive totals and cumulative totals in comparison with each other, using the same two axes. Although the method of depiction is versatile, the usual contexts in which this kind of chart is used are (1) industrial production and (2) marketing in order to compare monthly figures of either production or sales with (a) totals for a calendar year to date and (b) totals of all months of the most recent twelve-month period including the month concerned.

Example
The illustration below depicts the sales of a product. If sales had been constant, the lines joining the monthly plottings would show an exact letter Z, hence the name given to the chart. In this case a Z-like formation is partly discernible. The letters J, F, M ... D indicate the months of the calendar year to which the three values – (i) individual value, (ii) total for the calendar year to date and (iii) total for the previous 12 months to date – are all applicable.

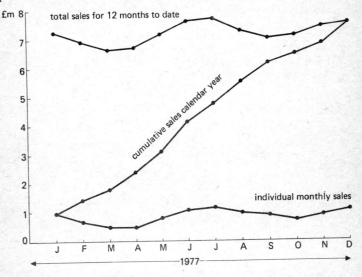

Zones. Equal-sized groups of elements or strata. In various sampling schemes, particularly in SYSTEMATIC SAMPLING, the population is divided into implicit strata or zones of equal size as a preliminary to the sample selection. One of the advantages of the use of zones is that if the sample is selected with equal probabilities the sample sizes in the zones are equal. This means that the simple formulae for REPLICATED SAMPLING are applicable.

Guide to Further Reading

This list of texts is divided into four sections. Section 1 provides some details of introductory texts, giving where relevant their major strengths. In section 2 a number of intermediate texts are classified in respect of a specific application or content. Section 3 contains some advanced expositions of mathematical statistics that will be useful for continued study of the subject, and Section 4 lists other useful publications, e.g. dictionaries and tables.

1 Introductory texts

J. K. Backhouse, *Statistics: An Introduction to Tests of Significance*, Longmans, 1967. This is a small and simple exposition of the subject, with useful chapters on probability, the chi-squared distribution, and a clear exposition of analysis of variance.

J. E. Freund, *Modern Elementary Statistics*, Prentice-Hall, 1974. J. E. Freund, *Statistics: A First Course*, Prentice-Hall, 1976. Both these books serve similar functions, but with slight differences of emphasis.

P. G. Hoel, *Elementary Statistics* (Probability and Mathematical Statistics Series), Wiley, 1976. This book was published in 1960 originally and has undergone several very successful editions. It is one of the few elementary books that tries to steer a middle course between dogmatic statement and detailed explanation of statistical formulae.

D. Huff, *How to Lie with Statistics*, Penguin, 1973. This book pays most attention to very simple concepts, and to the problems of graphical presentation. Its great merit is the ease with which it can be read.

M. J. Moroney, *Facts from Figures*, Penguin, 1969. Although this is listed as an introductory text because of its ease of exposition and its clarity, it progresses further into the theory of statistics than most of the books in this section, providing the reader with a simple exposition of analysis of covariance, and giving applications of other useful statistical tools.

J. Murdoch and J. A. Barnes, *Statistics: Problems and Solutions*, Macmillan, 1972. A useful practical text containing graded questions.

F. R. Oliver, *What Do Statistics Show?* Hodder and Stoughton, 1964. A clear elementary text, dealing primarily with descriptive statistics, derived statistics, index numbers, confidence intervals and sampling, but discussing most other basic concepts and analytical techniques.

R. E. Walpole, *Elementary Statistical Concepts*, Macmillan, 1976. R. E. Walpole, *Introduction to Statistics*, Collier-Macmillan, 1974. These books cover a syllabus similar to that covered by Freund above, but are written for the social scientist rather than for the business studies student. The first book is particularly valuable because of its concise treatment of non-parametric statistics.

2 Texts on specific topics and specific applications

Bayesian statistics

S. A. Schmidt, *Measuring Uncertainty: An Introduction to Bayesian Statistics*, Addison-Wesley, 1969.

Linear regression analysis

N. R. Draper, and H. Smith, *Applied Regression Analysis*, Wiley, 1966. An excellent treatment of the subject, probably the best text on the subject.

G. A. F. Seber, *Linear Regression Analysis*, Wiley, 1977. One of the most comprehensive of advanced texts on the subject, with useful appendices on matrix algebra, and orthogonal projections.

Non-parametric statistics

E. L. Lehmann, *Non-Parametrics: Statistical Methods Based on Ranks*, Holden-Day, 1975. Although this is a book about applications of non-parametric statistics, designed for a reader with only an elementary knowledge of probability, it satisfies the criterion of being one of the best theoretical texts on the subject not only by providing theoretical appendices, but by stating and proving the case for the power and efficiency of non-parametric methods.

R. J. Langley, *Practical Statistics for Non-mathematical People*, David and Charles, 1971. The writer's main concern is that of teaching statistics to medical students, but the greater proportion of the book is, in fact, devoted to non-parametric statistics. It discusses applications rather than theory.

S. Siegel, *Non-parametric Statistics for the Behavioural Sciences*, MacGraw-Hill, 1956. A well-organized book that explains most non-parametric tests, containing a valuable cover-inset classification of the tests and their applications.

Statistical inference

S. D. Silvey, *Statistical Inference*, Penguin, 1970. A brief but comprehensive text on the subject.

Survey sampling

W. G. Cochran, *Sampling Techniques*, Wiley, 3rd edn, 1977. A clear
and elegant book which presents a systematic treatment of sampling
techniques. The technical level is too high for non-statisticians.

L. Kish, *Survey Sampling,* Wiley, 1965. The outstanding textbook on
practical sampling principles and procedures. The treatment is
sufficiently non-technical to be intelligible to non-statisticians.

C. A. Moser and G. Kalton, *Survey Methods in Social Investigation*,
Heinemann, 1971. An excellent basic book on survey methodology.
Strongly recommended reading for all potential survey practitioners.

C. A. O'Muircheartaigh and C. D. Payne, (eds.), *The Analysis of Survey Data*, vol. 1: *Exploring Data Structures*; vol. 2: *Model-fitting*.
(Wiley: 1977) A broadly non-technical treatment of methods of data
analysis suitable for survey practitioners and graduate students in the
social sciences.

Business studies and accounting

J. P. Dickinson, *Statistics for Business, Finance and Accounting*, Mac-
donald and Evans, 1976. An introductory text written specially for
business studies students, but with some material (e.g. on non-
parametric statistics) not found in most business statistics introductory texts.

J. E. Freund and F. J. Williams, *Elementary Business Statistics*,
Prentice-Hall, 1974. D. Gregory and H. Ward, *Statistics for Business
Studies*, McGraw-Hill, 1974. These two books are similar in syllabus
content, applications and objectives.

P. G. Moore, *Principles of Statistical Techniques*, Cambridge University Press, 1969. A text used often in teaching statistics for business
studies degrees because of its clarity when dealing with fundamental
concepts.

D. Pitt Francis, *Statistical Method for Accounting Students*,
Heinemann, 1978. A statistical text specifically written for accounting students, using examples chosen mainly from accounting data.

T. H. and R. J. Wonnacott, *Introductory Statistics for Business and
Economics*, Wiley, 1976. This is much more than an 'introductory'
text, for it has expanded material on Monte Carlo sampling and on
Bayesian estimation.

Economics and econometrics

R. J. Allard, *An Approach to Econometrics*, Philip Allan, 1974. A
clear approach to a basic understanding of econometrics, dealing
extensively with linear regression analysis and having a special chapter on residuals, with numerous exercises at the end of each chapter.

R. E. Beals, *Statistics for Economists: An Introduction*, McNally, 1972.

C. R. Frank, *Statistics and Econometrics*, Holt, Rinehart and Winston, 1971. An extended treatment of both subjects, showing the relationship between the two, and having sections on descriptive statistics, probability, inference, linear regression and econometrics.

J. D. Hey, *Statistics in Economics*, Martin Robertson, 1974. A useful conceptual book on application of statistics to economics, requiring very little prior knowledge of mathematics.

A. C. and D. G. Mayes, *Introductory Economic Statistics*, Wiley, 1976. Another useful text that has the additional feature of numerous applications and extensive answer notes.

Educational and psychological

H. E. Garrett, *Statistics in Psychology and Education*, Longmans 1966. J. P. Guildford, *Statistics in Psychology and Education* McGraw-Hill, 1965. Texts dealing with ways in which statistics can be applied to education and the behavioural sciences.

Medical Statistics

W. M. Castle, *Statistics in Small Doses*, Churchill Livingstone, 1977.

Sir Austin Bradford Hill, *Principles of Medical Statistics*, Lancet, 1971.

H. O. Lancaster, *Introduction to Medical Statistics*, (Probability and Mathematical Statistics Series), Wiley, 1974.

Physical sciences and engineering

K. A. Brownlee, *Statistical Theory and Methodology in Science and Engineering*, Wiley, 1965.

Social statistics and demography

H. M. Blalock, *Social Statistics*, McGraw-Hill, 1972. Probably the best introductory textbook for social scientists. The mathematical level is fairly high but the treatment is clear and comprehensible.

F. Conway, *Sampling: An Introduction for Social Scientists*, Allen and Unwin, 1976.

P. R. Cox, *Demography, Cambridge University Press*, 1976.

B. Edwards, *Sources of Social Statistics*, Heinemann, 1974.

A. G. Johnson, *Social Statistics without Tears*, McGraw-Hill, 1977.

K. A. Yeomans, *Statistics for the Social Scientist*, Penguin, 1968.

3 Advanced texts for further study

Sir M. Kendall and A. Stuart, *The Advanced Theory of Statistics*, Griffin, 1976. A three-volume work spanning the whole of statistical theory, and a treatment of distribution theory, inference and relationship and design analysis.

G. U. Yule and Sir M. G. Kendall, *An Introduction to the Theory of Statistics*, Griffin, 1965. An older work providing much of the foundation for the study of the later three-volume work.

4 Dictionary and tables

Sir M. G. Kendall and W. Buckland, *Dictionary of Statistics*, Longmans/International Statistical Institute, 1975. A comprehensive encyclopaedia of all terms used in mathematical statistics.

J. Murdoch and J. A. Barnes. *Statistical Tables*, Macmillan, 1971. This is the most commonly used set of statistical tables. Other useful sets are the Cambridge Statistical Tables, and those published by Cassell and Heinemann. One of the largest is that published by the Institute of Mathematical Statistics: *Selected Tables in Mathematical Statistics*, American Mathematical Society, Vols I-III, 1972/75.